Social Media and Youth: Navigating the Digital Landscape

Swatantra Bahadur

Published by Swatantra Bahadur, 2023.

SOCIAL MEDIA AND YOUTH: NAVIGATING THE DIGITAL LANDSCAPE

First edition. December 20, 2023.

Copyright © 2023 Swatantra Bahadur.

ISBN: 979-8215427149

Written by Swatantra Bahadur.

Disclaimer

The information presented in this book is intended for general informational purposes only and should not be relied upon as a substitute for professional advice or judgment. The author and publisher are not responsible for any action taken by readers based on the information provided in this book. Readers should seek appropriate professional advice or conduct their own research before making decisions related to the topics discussed in this book. The views expressed in this book are those of the author and do not necessarily reflect the views of the publisher.

Contents

Introduction

In the ever-evolving tapestry of the digital age, social media has woven itself into the very fabric of our daily lives. Nowhere is this influence more palpable than in the lives of today's youth, who navigate a complex and dynamic landscape shaped by tweets, snaps, likes, and shares. The intersection of social media and the youth demographic is both profound and intricate, offering unprecedented opportunities for connection, expression, and education, but also presenting unique challenges and risks. This book delves into the multifaceted relationship between social media and the youth of today, exploring the transformative impact of these digital platforms on their social, psychological, and educational spheres. As we embark on this exploration, it becomes imperative to dissect the evolution of social media, its positive contributions to youth culture, as well as the darker undercurrents that demand our attention and understanding. In the pages that follow, we will journey through the historical roots of social media, tracing its trajectory from the early days of online forums to the sophisticated networks that define contemporary communication. We will examine the positive aspects, where social media serves as a catalyst for creativity, connectivity, and social change. Simultaneously, we will confront the negative facets, such as the rising tide of cyberbullying, the potential for addiction, and the pervasive impact of misinformation. Crucially, this exploration is not solely an analytical endeavor; it is a journey into the lived experiences of the youth who are both the architects and subjects of this digital revolution. Through case studies, personal narratives, and expert insights, we aim to provide a nuanced understanding of the intricate relationship between social media and the youth population. As we embark on this journey, we invite readers to critically engage with the subject matter, to reflect on the implications of social media on the youth of today, and to consider the role of parents, educators, and society at large in guiding the next generation through the challenges and opportunities of the digital age. The pages ahead are not just a chronicle of the impact of social media on youth; they are a call to action, a roadmap for fostering a healthy, responsible, and empowered digital future for the leaders of tomorrow.

A. Definition and significance of social media

In the landscape of modern communication, social media stands as a defining and transformative force. At its core, social media encompasses a diverse array of online platforms and technologies that facilitate the creation, sharing, and exchange of information, ideas, and multimedia content within virtual communities. These platforms range from well-known giants like Facebook, Twitter, Instagram, and Snapchat to emerging channels that continually shape the digital terrain.

The significance of social media is profound, permeating every facet of contemporary life. It has redefined the way individuals connect, communicate, and express themselves on a global scale. No longer confined by geographical boundaries, social media enables instantaneous communication and the formation of communities based on shared interests, experiences, or affiliations.

The democratization of information is a hallmark of social media, granting individuals the power to broadcast their thoughts, opinions, and experiences to a vast audience. This democratization, while empowering, also raises questions about the authenticity and reliability of information circulating in these digital spaces. The viral nature of content dissemination on social media platforms has the potential to influence public discourse, shape opinions, and even catalyze social movements.

Beyond personal interactions, social media plays a pivotal role in the realms of business, politics, education, and culture. It serves as a marketing tool for businesses, a platform for political discourse and activism, a space for learning and sharing knowledge, and a medium for cultural expression. The interconnectedness fostered by social media has given rise to a digital ecosystem where trends, ideas, and movements can rapidly gain momentum and influence.

However, this digital interconnectedness also poses challenges. The line between public and private life becomes blurred, raising concerns about privacy and the potential for exploitation. The addictive nature of social media engagement, coupled with the pressure to curate an idealized online persona, can impact mental health and well-being, especially among the youth.

In navigating the digital landscape shaped by social media, it becomes essential to critically examine its definition and significance. This examination lays the foundation for understanding the far-reaching impact of social media on individuals, communities, and societies, providing insights into both its positive contributions and the challenges it poses.

B. The youth demographic and their digital presence

The youth demographic, often referred to as the "digital natives," occupies a central and dynamic space within the vast expanse of social media. Defined by those typically in their teens to early twenties, this generation has grown up in a world where digital connectivity is not a luxury but an integral aspect of daily life. Understanding their digital presence on social media involves exploring the unique ways in which they interact, communicate, and construct their identities in the virtual realm.

Digital Native Identity:

- The youth demographic is characterized by their inherent fluency in digital technologies. For them, social media is not just a tool; it is an extension of their identity and a fundamental aspect of their social existence.

Communication and Connectivity:

- Social media serves as a primary mode of communication for young individuals, offering instant messaging, video calls, and multimedia sharing. This immediacy and accessibility redefine how they form and maintain relationships, both locally and globally.

Self-Expression and Identity Formation:

- Social media platforms provide a canvas for self-expression, allowing young users to curate their online personas. From carefully crafted profiles to sharing personal achievements, struggles, and aspirations, social media becomes a medium for shaping and expressing their identities.

Peer Influence and Social Validation:

• The need for social validation is amplified in the digital space. Likes, comments, and shares contribute to a sense of acceptance and popularity. This dynamic can influence behavior, preferences, and even self-worth, creating both positive and negative feedback loops.

Activism and Social Change:

• Social media empowers the youth to engage in social and political issues. Activism takes on new forms as young individuals use digital platforms to raise awareness, organize movements, and advocate for change. Hashtags become rallying cries, and online spaces transform into arenas for social and political discourse.

Challenges of Digital Presence:

• The constant connectivity and the pressure to maintain a curated online presence can lead to challenges such as cyberbullying, digital fatigue, and mental health concerns. Striking a balance between the digital and physical realms becomes crucial for overall well-being.

Educational Impact:

• Social media is not only a social space but also a source of information and learning. The youth use platforms for educational purposes, accessing news, tutorials, and online courses. However, discerning between credible and misleading information becomes a critical skill in this digital landscape.

Understanding the youth demographic and their digital presence is pivotal for grasping the broader implications of social media. It sheds light on the ways in which these digital natives navigate and contribute to the evolving digital culture, shaping the future of communication, relationships, and societal dynamics.

C. Purpose and scope of the book

The purpose of this book is to provide a comprehensive and insightful exploration of the intricate relationship between social media and the youth demographic. By delving into the multifaceted impact of digital platforms on the lives of young individuals, the book seeks to illuminate both the positive contributions and challenges posed by the pervasive influence of social media. Through in-depth analysis, real-world examples, and expert perspectives, the aim is to equip readers with a nuanced understanding of this complex intersection and foster meaningful conversations around the implications for individuals, communities, and society as a whole.

Scope of the Book:

Historical Evolution of Social Media:

• The book will trace the historical development of social media, providing context for the current digital landscape and highlighting key milestones that have shaped the interaction between youth and online platforms.

Positive Impacts:

• Exploring the positive aspects of social media on youth, the book will delve into how these platforms foster social connectivity, provide educational opportunities, facilitate creative expression, empower entrepreneurship, and serve as catalysts for activism and social change.

Negative Impacts:

• Addressing the challenges and risks, the book will examine the negative facets of social media, including mental health implications, cyberbullying, privacy concerns, addiction, and the spread of misinformation. Real-life case studies will illustrate the real-world consequences of these issues.

Role of Parents and Educators:

• Recognizing the vital role of parents and educators, the book will provide practical guidance on fostering digital literacy, establishing healthy digital habits, and promoting open communication to mitigate the potential negative impacts of social media on youth.

Case Studies and Personal Stories:

• The inclusion of case studies and personal narratives will humanize the discussion, offering readers a glimpse into the lived experiences of youth in the digital age. Success stories, cautionary tales, and insights from experts will enrich the narrative.

Strategies for Mitigation:

• The book will present strategies and recommendations for mitigating the negative impacts of social media, focusing on promoting digital well-being, building resilience, addressing online safety and privacy concerns, and encouraging positive online behaviors.

Future Trends and Challenges:

• Examining the trajectory of social media, the book will explore emerging trends, technologies, and potential challenges on the horizon. Ethical considerations and the role of regulation in shaping the future of social media will be discussed.

Conclusion and Call to Action:

• The book will conclude by summarizing key insights and calling for a proactive and informed approach to navigating the evolving landscape of social media and youth. It will emphasize the collective responsibility of individuals, communities, and society in fostering a healthy and empowered digital future.

By addressing both the positive and negative dimensions of social media's impact on youth, this book aims to contribute to a well-rounded understanding of this critical intersection, fostering dialogue and inspiring informed action in the face of the ever-evolving digital landscape.

II. The Evolution of Social Media

In the blink of an eye on the historical timeline, we find ourselves immersed in a digital era where the contours of communication and connectivity are defined by the ever-expanding realm of social media. The evolution of these online platforms mirrors a dynamic journey, from the humble beginnings of text-based forums to the immersive and interactive landscapes that captivate the attention of today's youth.

1. Birth of Online Communities:

- The genesis of social media can be traced back to the early days of online communities and forums. Bulletin board systems and Usenet paved the way for digital conversations, setting the stage for the interconnected web we know today.

2. Emergence of Early Social Networks:

- The late 1990s and early 2000s witnessed the rise of the first-generation social networks. Platforms like Friendster and MySpace introduced the concept of personal profiles and friend connections, providing users with novel ways to express their identities and connect with others.

3. The Facebook Revolution:

- In 2004, Mark Zuckerberg and his Harvard roommates launched Facebook, a platform initially designed for college students to connect. The success and rapid expansion of Facebook marked a turning point, introducing the term "social networking" to the global lexicon.

4. Microblogging and Real-Time Interaction:

• The advent of microblogging platforms like Twitter in 2006 brought a new dimension to online communication. Real-time updates, short-form content, and hashtags transformed the way information was shared, making it a powerful tool for both personal expression and breaking news.

5. Visual Storytelling:

• The 2010s witnessed a shift towards visual-centric platforms. Instagram, with its focus on photo and video sharing, and Snapchat, with ephemeral content, catered to the youth's desire for visual storytelling, creativity, and instantaneous expression.

6. Rise of Influencers and Content Creators:

• The democratization of content creation became evident with the rise of YouTube and platforms like Vine (later succeeded by TikTok). A new generation of influencers and content creators emerged, shaping trends, pop culture, and redefining notions of fame.

7. Mobile Dominance and App Ecosystems:

• The ubiquity of smartphones further accelerated the evolution of social media. Apps became the primary gateway to digital interaction, fostering an ecosystem where platforms competed for attention, innovation, and user engagement.

8. Virtual Realities and Beyond:

• As we step into the current era, the integration of virtual and augmented reality technologies is reshaping social media. Platforms explore immersive experiences, interactive environments, and new ways for users to engage beyond the confines of traditional screens.

9. Challenges and Reflections:

• Alongside the evolution of social media, challenges have emerged. Issues of privacy, mental health implications, and the spread of misinformation prompt a collective reflection on the impact of these digital landscapes on the youth who navigate them.

The evolution of social media is a testament to the rapid pace of technological advancement and the profound influence these platforms wield in shaping the experiences of today's youth. Understanding this evolution provides a crucial lens through which we can unravel the complex dynamics of social media's impact on the lives of the digital generation. As we traverse this ever-changing digital landscape, the evolution of social media stands as a compelling narrative, reflecting not only technological progress but also the evolving social fabric of our interconnected world.

A. HISTORICAL DEVELOPMENT of social media platforms

Pre-Internet Precursors (Early 1970s - 1990s):

• The roots of social media can be traced back to early computer-based communication systems. Bulletin board systems (BBS) and online forums, such as CompuServe (1969) and The WELL (1985), laid the groundwork for digital discussions and community building.

Six Degrees and the Birth of Social Networking (1997):

• In 1997, the platform "Six Degrees" marked a significant milestone as the first recognizable social networking site. Users could create profiles and connect with friends, laying the foundation for the interconnected online social experience.

Friendster and the Dawn of Social Networks (2002):

• Friendster (2002) popularized the concept of social networking by allowing users to connect with friends, share content, and discover new connections. While it faced challenges, it set the stage for future platforms by demonstrating the social appeal of online networks.

MySpace and the Rise of User-Generated Content (2003):

• MySpace (2003) emerged as a cultural phenomenon, enabling users to customize profiles with music, images, and layouts. It became a hub for musicians, artists, and individuals seeking self-expression, showcasing the potential of user-generated content.

Facebook: From Campus to Global Dominance (2004):

• Founded in 2004 by Mark Zuckerberg, Facebook initially targeted college students. Its rapid expansion and introduction of the News Feed feature in 2006 transformed social media into a dynamic and real-time platform, setting the stage for its global dominance.

Twitter and Microblogging (2006):

• Twitter, launched in 2006, introduced microblogging, limiting posts to 140 characters. This concise format facilitated real-time communication, breaking news, and hashtag activism, contributing to the emergence of a new form of online discourse.

Visual Revolution with Instagram (2010):

• Instagram's arrival in 2010 shifted the focus to visual content. Its emphasis on photo and video sharing, along with creative filters, appealed to users seeking a more visually-driven and streamlined social experience.

Snapchat's Ephemeral Content (2011):

• Snapchat disrupted the social media landscape in 2011 by introducing ephemeral content. Disappearing messages and Stories appealed to younger users, emphasizing spontaneity and reducing the permanence of online interactions.

Video Dominance: YouTube and TikTok (2005, 2016):

• YouTube's launch in 2005 popularized video content on social media. TikTok, introduced in 2016, embraced short-form videos and user-generated content, redefining how users engage with and create online content.

Professional Networking with LinkedIn (2003) and Growth of Niche Platforms:

• LinkedIn, founded in 2003, focused on professional networking, showcasing the diversification of social media purposes. Niche platforms like Pinterest (2010) and Reddit (2005) highlighted the trend of catering to specific interests and communities.

Current Landscape and Future Trends:

• The current social media landscape reflects a diverse array of platforms, each catering to unique preferences and demographics. Emerging trends include the integration of virtual reality, augmented reality, and the ongoing evolution of platform ecosystems.

Understanding the historical development of social media platforms unveils a captivating narrative of innovation, cultural shifts, and the relentless pursuit of connectivity. From rudimentary online forums to the immersive experiences of today, social media platforms have evolved into powerful influencers, shaping the way individuals communicate, express themselves, and navigate the digital age.

B. The rise of social media in youth culture

Pioneering a New Era (Late 1990s - Early 2000s):

• As the 21st century dawned, the youth of the late 1990s and early 2000s found themselves at the forefront of a digital revolution. Friendster and MySpace emerged as pioneers, introducing the concept of online profiles and social connections. For the first time, young individuals could curate their digital identities and connect with peers beyond the confines of physical space.

Facebook: Campus Origins to Global Domination (2004-2010):

• The launch of Facebook in 2004, initially limited to college campuses, provided a virtual playground for youth culture to flourish. The platform's expansion beyond campuses and introduction of the News Feed fueled a cultural shift, transforming the way young people communicated, shared experiences, and maintained connections.

Twitter and the Rise of Microblogging (Late 2000s):

• Twitter, with its succinct 140-character format, became a cultural phenomenon. Young users embraced microblogging as a dynamic way to share thoughts, experiences, and engage in real-time conversations. Hashtags emerged as digital rallying cries, amplifying youth voices on a global scale.

Instagram and the Visual Revolution (2010s):

• The 2010s witnessed a visual renaissance with the rise of Instagram. Focused on photo and video sharing, the platform became a canvas for self-expression. Filters, Stories, and Explore features allowed youth to curate visually stunning narratives, fostering a culture of creativity and personal branding.

Snapchat and Ephemeral Content (2011-2015):

• Snapchat disrupted traditional notions of online permanence by introducing ephemeral content. The youth embraced disappearing messages and Stories as a form of unfiltered, real-time expression, reflecting a desire for authenticity and spontaneity in digital interactions.

YouTube and the Era of Content Creators (2000s - Present):

• YouTube transformed into a platform where young individuals could become content creators, sharing videos that resonated with global audiences. The rise of YouTubers and vloggers redefined celebrity culture, as authenticity and relatability became valued traits among youth influencers.

TikTok: Reshaping Creativity and Trends (2016-Present):

• TikTok, born in 2016, marked a paradigm shift in content creation. Short-form videos, lip-syncing, and challenges empowered youth to express themselves in new ways, contributing to the rapid creation and dissemination of trends. Its algorithm-driven discovery engine propelled ordinary users into viral stardom.

Influence on Language and Communication:

• Social media platforms introduced a lexicon that became an integral part of youth culture. Abbreviations, emojis, and meme culture evolved as unique forms of online expression, shaping the way young individuals communicated and connected with their peers.

Social Activism and Digital Advocacy:

• Youth culture on social media transcends personal expression; it is a powerful catalyst for social change. Hashtag movements, online petitions, and digital advocacy campaigns have empowered young activists to mobilize and amplify their voices on issues ranging from climate change to social justice.

Challenges and Nuances:

• As social media became woven into the fabric of youth culture, challenges surfaced. Issues of cyberbullying, digital addiction, and the pressure to maintain curated online personas raised concerns about the well-being of young users, prompting reflections on the impact of constant connectivity.

The rise of social media in youth culture is a dynamic narrative that intertwines with broader societal changes. From the early days of online profiles to the current era of influencer culture and digital activism, social media has not only reflected but also shaped the cultural preferences, communication styles, and societal values of the youth. It stands as a testament to the transformative power of digital platforms in shaping the very essence of how young individuals connect, communicate, and express themselves in the 21st century.

C. Key statistics and trends in youth social media usage

Global Penetration and Demographics:

• As of [latest year], over [percentage]% of the global youth population actively engages with social media platforms. The majority falls within the age range of 13 to 24, with [specific age group] representing the highest percentage of users.

Dominant Platforms:

• [Current year] sees [Platform A] and [Platform B] as the primary social media choices for youth. While [Platform A] continues to dominate in terms of overall user base, [Platform B] has experienced significant growth, particularly among [specific demographic].

Mobile-First Engagement:

• [Percentage]% of youth access social media exclusively through mobile devices. The shift to mobile-first engagement reflects the seamless integration of social media into the daily lives of young individuals, providing constant connectivity and instant access to digital content.

Daily Active Users (DAU):

• On average, youth users spend approximately [average hours] per day on social media platforms. This includes both intentional browsing and passive engagement, emphasizing the platforms' central role in shaping daily routines and leisure activities.

Visual Content Dominance:

• Visual content remains a driving force, with [percentage]% of youth favoring platforms that prioritize images and videos. This trend aligns with the rise of platforms like Instagram, TikTok, and Snapchat, showcasing the preference for visually-driven communication and storytelling.

Influencer Culture Impact:

• [Percentage]% of youth are influenced by content creators and influencers on social media. The rise of influencer culture has transformed these individuals into aspirational figures, shaping trends, consumer behavior, and even career aspirations among the youth demographic.

Shift to Private Messaging Apps:

• Private messaging apps, such as [App A] and [App B], have seen a [percentage]% increase in youth usage. This shift reflects a desire for more intimate and private communication, away from the public gaze of traditional social media platforms.

Ephemeral Content Preference:

• [Percentage]% of youth express a preference for platforms that offer ephemeral content features, such as disappearing messages and Stories. This trend aligns with the desire for spontaneous and authentic interactions, contributing to the popularity of platforms like Snapchat and Instagram Stories.

Concerns about Online Safety:

• [Percentage]% of youth express concerns about online safety and privacy on social media. Issues such as cyberbullying, data breaches, and unwanted attention impact the overall perception of social media, prompting calls for enhanced security measures and digital literacy initiatives.

Digital Well-being Initiatives:

• [Percentage]% of youth actively engage with digital well-being features on social media platforms. The integration of tools like screen time monitoring, notification controls, and mental health resources underscores a growing awareness of the potential negative impacts of excessive digital engagement.

These key statistics and trends underscore the profound influence of social media on the youth demographic. As digital natives continue to shape and reshape the landscape, staying attuned to these patterns is essential for understanding the evolving dynamics of youth social media usage and its broader implications for society.

III. Positive Impacts of Social Media on Youth

In the tapestry of contemporary youth culture, the advent of social media has woven threads of connection, empowerment, and positive transformation. Far from being merely a virtual playground, these digital platforms have emerged as powerful catalysts for growth, expression, and community-building among the youth demographic. Here, we delve into the positive impacts that social media imparts upon the lives of young individuals.

Social Connectivity and Peer Relationships:

• Social media acts as a virtual bridge, connecting young individuals with peers locally and globally. It provides a platform for forging new friendships, maintaining existing relationships, and fostering a sense of belonging and camaraderie.

Information and Educational Opportunities:

• Social media serves as a vast repository of knowledge, enabling youth to access educational resources, news, and diverse perspectives. Platforms like YouTube and Instagram offer tutorials, informative content, and opportunities for skill development, democratizing access to learning.

Creative Expression and Self-Identity:

• Youth leverage social media as a canvas for creative expression, showcasing talents in art, music, writing, and more. The ability to curate profiles allows them to shape and share their identities authentically, fostering a sense of individuality and self-discovery.

Entrepreneurship and Online Opportunities:

• Social media platforms provide a launchpad for young entrepreneurs and creatives. From launching small businesses on Instagram to monetizing content on YouTube, these platforms offer unprecedented opportunities for youth to showcase their skills, products, and services.

Cultural Exchange and Global Awareness:

• Social media acts as a global agora, facilitating cultural exchange and awareness. Youth engage in conversations with peers from diverse backgrounds, fostering understanding and appreciation for different cultures, perspectives, and societal issues.

Activism and Social Change:

• Social media emerges as a powerful tool for activism and advocacy. Youth-driven movements, propelled by hashtags and viral campaigns, amplify voices, raise awareness, and mobilize support for social and environmental causes, exemplifying the potential for positive societal impact.

Supportive Communities and Mental Health Advocacy:

• Online communities on platforms like Reddit and Facebook provide spaces for individuals facing similar challenges to connect and support each other. Social media also serves as a platform for mental health awareness, reducing stigma, and promoting positive discussions around well-being.

Access to Diverse Perspectives and Role Models:

• Social media exposes young individuals to a diverse array of perspectives, ideas, and role models. Influencers and content creators serve as inspirational figures, offering insights into various fields, fostering aspirations, and broadening horizons.

Networking and Professional Opportunities:

• Platforms like LinkedIn enable youth to build professional networks, connect with mentors, and discover career opportunities. Social media acts as a dynamic professional space, allowing young individuals to showcase their skills and accomplishments to a global audience.

Global Collaboration and Innovation:

• Social media facilitates collaboration beyond geographical boundaries. Youth engage in global projects, share innovations, and participate in cross-cultural initiatives, fostering a spirit of collaboration and innovation that transcends traditional constraints.

By embracing these positive impacts, social media empowers the youth to navigate the digital landscape with resilience, creativity, and a sense of social responsibility. As these digital natives continue to shape the future, the positive influences of social media on youth culture remain a testament to the transformative potential of online platforms.

A. Social connectivity and peer relationships

In the digital tapestry of social media, one of its most profound and transformative threads is the ability to foster social connectivity and nurture peer relationships. The impact of these platforms on the way young individuals form and maintain connections is nothing short of revolutionary, transcending geographical constraints and redefining the very essence of interpersonal relationships.

Global Village of Friendship:

• Social media serves as a global village, where youth can effortlessly connect with peers from different corners of the world. Platforms such as Facebook, Instagram, and Twitter become the virtual squares where cultural exchange, shared interests, and diverse perspectives converge, breaking down physical barriers to friendship.

Maintaining Long-Distance Relationships:

• For youth separated by geographical distances, social media becomes a lifeline for maintaining relationships. Through real-time communication, photo sharing, and video calls, these platforms bridge the gap, allowing friendships to flourish irrespective of physical proximity.

Diverse Social Circles:

• Social media platforms enable youth to create and engage with diverse social circles that may not be feasible in offline settings. This diversity fosters a rich tapestry of perspectives, experiences, and backgrounds, contributing to a more inclusive and globally aware youth culture.

Instantaneous Communication:

• The immediacy of social media communication transforms the way youth interact. Instant messaging, voice notes, and video calls provide an instantaneous and continuous channel for connection, allowing young individuals to share moments, thoughts, and experiences in real time.

Virtual Support Systems:

• Online platforms become virtual support systems where youth can lean on each other during both triumphs and tribulations. From offering words of encouragement to providing solace during challenging times, social media fosters a sense of emotional support and community.

Discovery of Like-Minded Communities:

• Social media enables the discovery of niche communities based on shared interests. Whether it's a love for a particular music genre, a passion for environmental activism, or an affinity for niche hobbies, these platforms facilitate the formation of like-minded communities that might be challenging to find locally.

Reducing Social Isolation:

• For those who might feel socially isolated offline, social media becomes an avenue for connection and belonging. Online friendships can be particularly empowering for individuals who may struggle with in-person social interactions, offering a space where they can express themselves freely.

Real-Time Event Participation:

• Youth connect and share experiences in real-time during events, whether they're local gatherings or global phenomena. Live-tweeting, sharing Stories, and participating in virtual events enhance the sense of shared experiences, reinforcing a feeling of collective participation.

Peer Support Networks:

• Social media acts as a platform for the creation of peer support networks, particularly in areas such as mental health. Online groups and communities provide spaces for youth to share experiences, seek advice, and offer empathy, contributing to a culture of mutual understanding and support.

Cultural Exchange and Understanding:

• Through social media, youth engage in cross-cultural friendships, promoting understanding and appreciation for diverse perspectives. Platforms facilitate the sharing of traditions, customs, and daily life, fostering a sense of global citizenship among the youth demographic.

In the realm of social connectivity and peer relationships, social media stands as a powerful facilitator, weaving a network that transcends physical boundaries and propels youth culture into a realm of interconnectedness, understanding, and shared experiences. It is through these digital threads that the fabric of youth relationships becomes not only richer but also more expansive, creating a global tapestry of friendships that redefine the very nature of social connection in the digital age.

B. Information and educational opportunities

In the dynamic landscape of social media, one of its transformative facets is the role it plays in providing information and educational opportunities to the youth. These platforms, far from being mere sources of entertainment, have evolved into robust repositories of knowledge, breaking down traditional barriers to learning and fostering a culture of continuous education among young individuals.

Access to Diverse Knowledge Sources:

• Social media platforms serve as gateways to an expansive world of information. Youth have unprecedented access to diverse knowledge sources, ranging from reputable news outlets and educational institutions to grassroots movements and independent content creators.

Interactive Learning Experiences:

• Platforms like YouTube and Instagram offer interactive learning experiences through tutorials, webinars, and educational content. The visual and interactive nature of these platforms transforms learning into a dynamic process, engaging the youth in a way that traditional educational formats may not.

Learning Beyond Classroom Walls:

• Social media extends the boundaries of the traditional classroom, allowing youth to learn beyond the confines of textbooks. Whether it's exploring historical events through immersive Instagram Stories or engaging in real-time discussions on Twitter, these platforms augment formal education.

Global Educational Communities:

• Online communities and groups on platforms like Facebook and Reddit create spaces for global educational exchange. Youth can seek advice, share resources, and engage in discussions with peers from around the world, fostering a sense of global awareness and collaboration.

Digital Literacy and Skills Development:

• Social media platforms contribute to the development of digital literacy skills essential for the modern era. Youth engage in content creation, learn about online safety, and navigate digital spaces, acquiring skills that are increasingly relevant in the digital landscape.

Online Courses and Tutorials:

• Social media serves as a hub for online courses and tutorials, providing youth with opportunities to enhance their skills or explore new interests. Platforms like LinkedIn Learning and TikTok's #LearnOnTikTok feature short, informative videos covering a wide array of subjects.

Inspiration and Aspirational Content:

• Influencers and content creators on platforms like Instagram and YouTube inspire youth by sharing their educational journeys and accomplishments. This aspirational content encourages young individuals to pursue their passions, fostering a culture of continuous learning and self-improvement.

Real-Time News and Current Events:

• Social media platforms act as real-time news sources, keeping youth informed about current events and global affairs. The immediacy of platforms like Twitter allows young individuals to stay abreast of developments as they unfold, contributing to a more informed and socially aware generation.

Citizen Journalism and Activism:

• Youth on social media engage in citizen journalism, reporting on local and global issues. Platforms become avenues for activism, enabling young individuals to raise awareness, share information, and mobilize support for social and environmental causes they are passionate about.

Educational Campaigns and Challenges:

• Educational campaigns and challenges on platforms like TikTok leverage the power of gamification to make learning fun and engaging. From science experiments to language challenges, these initiatives foster a positive and interactive approach to acquiring knowledge.

In the realm of information and education, social media emerges not only as a tool for dissemination but as a transformative force that empowers the youth to become active participants in their own learning journeys. By leveraging these platforms, young individuals navigate a digital landscape rich with educational opportunities, expanding their horizons and shaping a generation that values knowledge as a key driver of personal and societal growth.

C. CREATIVE EXPRESSION and self-identity

Within the digital realm of social media, a vibrant tapestry of creative expression and self-identity unfolds among the youth. These platforms, far more than just spaces for social interaction, have become dynamic canvases where young individuals paint the intricate brushstrokes of their identities and showcase their artistic prowess. Here, we explore the profound impact of social media on fostering creative expression and shaping self-identity among the youth.

Digital Storytelling through Visual Media:

• Social media platforms, particularly Instagram, Snapchat, and TikTok, have revolutionized the art of storytelling. Through captivating visuals, short videos, and creative edits, youth craft narratives that reflect their experiences, passions, and aspirations, shaping a unique digital story of self.

Photography and Visual Aesthetics:

• Platforms like Instagram celebrate the art of photography and visual aesthetics. Youth showcase their creativity through carefully curated feeds, experimenting with filters, compositions, and themes. Social media acts as a digital gallery where individuals express themselves visually and refine their artistic skills.

User-Generated Content and Collaboration:

• Social media platforms empower youth to become content creators. From YouTube vlogs to collaborative projects on TikTok, these spaces facilitate the creation and sharing of user-generated content, fostering a culture of collaboration and artistic exploration.

Virtual Art Galleries and Portfolios:

• For budding artists, platforms like Instagram and Pinterest serve as virtual galleries and portfolios. Youth share their drawings, paintings, digital art, and other creations, gaining exposure, feedback, and opportunities for collaboration within the global creative community.

Music and Creative Expression:

• Social media platforms have democratized music creation and sharing. Apps like SoundCloud and TikTok enable aspiring musicians to share their compositions, collaborate with other artists, and gain recognition, transforming social media into a stage for musical expression.

Personal Branding and Style Evolution:

• Youth use social media to cultivate personal brands, curating their online personas to reflect their unique styles and interests. Platforms like TikTok and Instagram Reels offer creative formats for self-expression, allowing individuals to showcase their personalities and fashion sense.

Writing and Microblogging:

• Microblogging platforms such as Twitter provide a space for youth to express themselves through written snippets. These concise expressions become a form of literary art, fostering a culture of concise and impactful storytelling within the constraints of character limits.

Experimental Video Content:

• Platforms like YouTube enable youth to experiment with long-form video content, from tutorials and documentaries to comedic sketches and video essays. This freedom to create diverse content allows for self-discovery and the exploration of various creative avenues.

Meme Culture and Humor Expression:

• Meme culture on platforms like Twitter and Reddit serves as a unique form of creative expression and humor. Youth engage in meme creation to comment on societal trends, share inside jokes, and participate in the evolving language of online humor.

Empowering LGBTQ+ Expression:

• Social media becomes a vital space for LGBTQ+ youth to express their identities, share experiences, and foster a sense of community. Platforms empower individuals to share their journeys, challenge stereotypes, and contribute to the broader conversation on diversity and inclusion.

In the realm of creative expression and self-identity, social media platforms emerge as not just canvases but as dynamic studios where the youth paint the masterpiece of their lives. These spaces transcend the limitations of traditional forms of expression, allowing young individuals to explore, experiment, and showcase their unique identities in a global gallery of creativity and self-discovery.

D.   Entrepreneurship   and   online opportunities

Within the expansive landscape of social media, a transformative wave of entrepreneurship has emerged, empowering youth to transform their passions into viable online ventures. These platforms, beyond being hubs of connection, have become fertile grounds for young individuals to cultivate entrepreneurial spirit, create businesses, and explore innovative opportunities. Here, we delve into the profound impact of social media on fostering entrepreneurship and providing diverse online opportunities for the youth.

E-Commerce and Digital Storefronts:

• Social media platforms, notably Instagram and Facebook, serve as dynamic digital storefronts for young entrepreneurs. Through features like Instagram Shops and Facebook Marketplace, youth can showcase and sell products directly to a global audience, breaking down traditional barriers to entry in the retail space.

Influencer Marketing and Brand Collaborations:

• Social media influencers have become entrepreneurs in their own right. Platforms like YouTube, Instagram, and TikTok offer opportunities for youth to collaborate with brands, monetize their influence, and build sustainable income streams through sponsored content and brand partnerships.

Content Monetization on YouTube and Patreon:

• Aspiring content creators on platforms like YouTube can leverage monetization features to earn income through ad revenue and channel memberships. Additionally, platforms like Patreon provide a space for fans to support creators directly, fostering a sustainable model for online entrepreneurship.

Dropshipping and Online Retail Innovation:

• The rise of dropshipping, facilitated by platforms like Shopify, enables youth to venture into e-commerce without the need for substantial upfront investments. Social media serves as a powerful marketing tool for dropshipping entrepreneurs, allowing them to reach a vast audience.

Freelancing and Gig Economy Platforms:

• Social media acts as a bridge to freelance opportunities through platforms like Upwork, Fiverr, and Freelancer. Youth can showcase their skills, offer services, and connect with clients globally, entering the digital gig economy and establishing themselves as freelancers or consultants.

Digital Marketing Agencies and Consultancies:

• Youth with expertise in digital marketing leverage social media to establish agencies and consultancies. Platforms like LinkedIn serve as professional spaces for networking, client acquisition, and showcasing the results of successful campaigns, paving the way for entrepreneurial ventures in the digital marketing sphere.

Online Courses and E-Learning Platforms:

• Social media facilitates the creation and promotion of online courses. Youth entrepreneurs can share their expertise, whether in photography, coding, or fitness, through platforms like Udemy or Skillshare, turning their knowledge into a valuable product.

App Development and Tech Startups:

• The app economy has thrived with young entrepreneurs leveraging social media to promote their applications. Platforms like Instagram and Twitter become crucial for building hype, engaging potential users, and securing early adopters for new tech startups.

Social Media Management and Consulting:

• Proficient users of social media platforms often turn their skills into entrepreneurship opportunities by offering social media management services. Youth can establish their consulting businesses, assisting individuals and businesses in navigating the complexities of online presence and branding.

Crowdfunding and Community Support:

• Social media provides a launchpad for crowdfunding campaigns through platforms like Kickstarter and Indiegogo. Youth entrepreneurs can present their innovative ideas to a global audience, garnering support and funding from a community of backers who resonate with their vision.

In the expansive realm of entrepreneurship and online opportunities, social media emerges as a dynamic force that not only connects young innovators with global audiences but also democratizes access to markets and resources. These platforms empower youth to realize their entrepreneurial dreams, fostering a culture of innovation, resilience, and limitless possibilities in the digital age.

E. Activism and social change

Within the digital landscape of social media, a powerful current of activism and social change flows through the veins of youth culture. These platforms, far from being mere spaces for social interaction, have become catalysts for movements, providing young individuals with the tools to raise their voices, challenge norms, and advocate for meaningful societal transformations. Here, we explore the profound impact of social media on activism and its role in driving positive social change among the youth.

Hashtag Movements as Catalysts:

• Social media platforms, particularly Twitter and Instagram, have birthed and amplified countless hashtag movements. From

#BlackLivesMatter to #MeToo, these movements transcend digital boundaries, becoming powerful vehicles for youth to address social injustices, share personal stories, and mobilize for change.

Amplifying Marginalized Voices:

• Social media provides a platform for marginalized voices that may be underrepresented in traditional media. Youth activists leverage these platforms to shed light on issues affecting their communities, fostering inclusivity, and challenging systemic inequalities.

Digital Advocacy and Awareness Campaigns:

• Youth-led digital advocacy campaigns on platforms like Change.org and Avaaz allow individuals to mobilize support for a myriad of causes. Social media becomes a tool for raising awareness, garnering signatures for petitions, and driving public sentiment towards specific issues.

Live Streaming for Real-Time Impact:

• Live streaming features on platforms like Facebook and Instagram provide real-time coverage of events, protests, and activism. Youth activists use this immediacy to document and share unfolding situations, providing a raw and unfiltered perspective to their audience.

Educational Initiatives and Information Dissemination:

• Social media acts as an educational hub for spreading information about social issues and injustices. Youth activists curate content, infographics, and articles to inform and educate their peers, fostering a generation that is well-informed and socially conscious.

Global Solidarity and Collaborative Movements:

• Platforms like Twitter enable youth activists to connect with like-minded individuals globally. This interconnectedness fosters collaborative movements, where activists from different parts of the world unite to address shared concerns and advocate for global change.

Rapid Response and Mobilization:

• Social media facilitates rapid response to emerging issues. Youth activists can swiftly mobilize their followers, organizing protests, boycotts, and other forms of direct action in response to pressing social and political events.

Political Engagement and Voter Mobilization:

• Youth activists use social media to engage their peers in political discourse, encourage voter registration, and mobilize for elections. Platforms become spaces for discussing policy issues, holding politicians accountable, and promoting civic participation.

Environmental Activism and Sustainability Advocacy:

• Youth-led environmental movements, such as Fridays for Future, leverage social media to advocate for climate action. Activists share information on environmental issues, organize virtual strikes, and encourage sustainable practices, fostering a global movement for a greener future.

Digital Narratives for Social Justice:

• Social media becomes a storytelling platform for social justice narratives. Youth activists share personal stories, experiences, and perspectives, creating a collective narrative that challenges societal norms, prejudices, and stereotypes.

In the realm of activism and social change, social media emerges as a powerful force that democratizes advocacy, amplifies voices, and catalyzes movements for a more just and equitable world. It empowers the youth to be agents of change, offering a digital megaphone that transcends geographical barriers and transforms online conversations into real-world impact.

IV. Negative Impacts of Social Media on Youth

While social media has undoubtedly transformed the landscape of communication and connection, it carries with it a shadow side that casts concerns over the well-being of the youth. As these platforms become integral to daily life, a range of negative impacts has surfaced, influencing the mental, emotional, and social dimensions of young individuals. Here, we delve into the darker facets of social media and its potential adverse effects on the youth.

Cyberbullying and Online Harassment:

• Social media provides a breeding ground for cyberbullying, where youth may face harassment, threats, or the spread of false information. The anonymity afforded by online platforms can amplify the impact of such negative behaviors, leading to lasting emotional and psychological consequences.

Body Image Concerns and Unrealistic Standards:

• The prevalence of curated, filtered, and idealized images on platforms like Instagram can contribute to body image issues among the youth. Constant exposure to unrealistic beauty standards may lead to feelings of inadequacy, low self-esteem, and a pursuit of unattainable ideals.

Social Comparison and FOMO (Fear of Missing Out):

• The curated nature of social media profiles often fosters an environment of social comparison. Youth may experience FOMO, feeling left out or inadequate when exposed to the seemingly perfect lives and experiences of their peers, contributing to anxiety and depression.

Digital Addiction and Impaired Concentration:

• Excessive use of social media has been linked to digital addiction, impacting the ability of youth to concentrate and engage in offline activities. Constant notifications, scrolling, and the need for online validation may lead to diminished attention spans and academic performance.

Privacy Concerns and Data Exploitation:

• Youth may unknowingly expose themselves to privacy risks as they share personal information on social media. The collection and exploitation of user data by third parties raise concerns about online safety, identity theft, and the potential for targeted advertising that can manipulate consumer behavior.

Sleep Disturbances and Blue Light Exposure:

• Late-night social media use, often facilitated by smartphones, can lead to sleep disturbances. The exposure to blue light emitted by screens interferes with circadian rhythms, contributing to difficulties falling asleep and impacting overall sleep quality among the youth.

Negative Impact on Mental Health:

• Research suggests a correlation between heavy social media use and mental health issues, including depression, anxiety, and feelings of loneliness. Constant exposure to curated content, online drama, and the pressure to conform to digital norms may contribute to mental health challenges.

Online Predators and Exploitation:

• Social media platforms may become hunting grounds for online predators seeking to exploit vulnerable youth. The ease of creating fake profiles and the potential for unsolicited contact pose risks to the safety and well-being of young individuals online.

Echo Chambers and Polarization:

• Algorithms on social media platforms often contribute to the creation of echo chambers, where users are exposed to content that aligns with their existing beliefs. This can lead to increased polarization, limiting exposure to diverse perspectives and contributing to societal divisions.

Impact on Real-Life Relationships:

• Excessive use of social media may strain real-life relationships, as individuals prioritize online interactions over face-to-face communication. The constant need for validation and comparison may lead to feelings of disconnection and interpersonal challenges in offline settings.

While social media offers unparalleled connectivity, it is crucial to navigate its landscapes with awareness of the potential negative impacts. Balancing digital engagement with offline well-being becomes imperative to ensure that the youth harness the benefits of social media while mitigating its darker consequences on mental, emotional, and social dimensions.

A. MENTAL HEALTH CHALLENGES and social comparison

The digital era, epitomized by the ubiquity of social media, has ushered in a complex interplay between online interactions and mental well-being, particularly among the youth. Amidst the seemingly endless scroll of curated content, a tapestry of mental health challenges and social comparison unravels, casting shadows on the psychological landscape of young individuals navigating the digital realm.

Constant Social Comparison:

• Social media platforms, with their curated profiles and highlight reels, foster an environment ripe for social comparison. The constant exposure to the achievements, lifestyles, and perceived happiness of peers can evoke feelings of inadequacy and a distorted sense of self-worth.

FOMO and Anxiety:

• The Fear of Missing Out (FOMO) is a pervasive emotion driven by the fear of being excluded from social activities or experiences shared on social media. Youth experiencing FOMO may grapple with anxiety, feeling compelled to constantly engage with online content to avoid the fear of being left out.

Body Image Issues:

• The prevalence of meticulously edited and filtered images on platforms like Instagram contributes to body image concerns among the youth. Exposure to idealized beauty standards may lead to self-esteem issues, unhealthy dieting behaviors, and a distorted perception of one's own body.

Validation Seeking Behavior:

• The pursuit of likes, comments, and followers on social media can become a relentless quest for validation. Youth may tie their self-worth to online approval, leading to a constant need for affirmation that can be emotionally draining and contribute to anxiety.

Online Bullying and Cyberbullying:

• Social media platforms provide a space where cyberbullying can thrive. Youth who experience online harassment may face heightened levels of stress, anxiety, and even depression. The persistent nature of digital communication can exacerbate the impact of such negative interactions.

Digital Exhaustion and Overstimulation:

• The incessant flow of information, notifications, and content on social media contributes to digital exhaustion. Youth may experience feelings of overwhelm, fatigue, and difficulty in disconnecting from the digital world, impacting their overall mental well-being.

Depression and Social Isolation:

• Excessive use of social media, coupled with the negative emotions stirred by constant social comparison, has been linked to symptoms of depression. Youth may retreat into social isolation, finding solace in the digital world while disengaging from real-life social interactions.

Impacts on Sleep Patterns:

• Late-night social media use, driven by the desire to stay connected, can disrupt sleep patterns. Exposure to screens before bedtime, coupled with the emotional impact of online interactions, may contribute to sleep disturbances and exacerbate mental health challenges.

Pressure to Conform and Fit In:

• The need to fit in with online trends and conform to digital norms can create immense pressure on youth. The fear of being judged or excluded may lead to a constant effort to mold one's online persona, contributing to heightened stress levels and a sense of inauthenticity.

Erosion of Real-Life Connections:

• Excessive engagement with social media may erode real-life connections. The focus on online relationships can lead to a sense of disconnection from the immediate physical environment, impacting the quality of face-to-face interactions and contributing to feelings of loneliness.

In the intricate dance between social media and mental health, the youth find themselves grappling with a myriad of challenges. Navigating the complexities of social comparison requires a nuanced understanding of the digital landscape, emphasizing the importance of fostering a healthy relationship with technology, promoting digital literacy, and nurturing offline well-being.

B. Cyberbullying and online harassment

As the digital age continues to evolve, the rise of cyberbullying and online harassment casts a shadow on the seemingly boundless realms of social media. For the youth, these platforms, while fostering connectivity, can also become arenas of adversity, where the impact of harmful online behaviors reverberates through the intricate threads of their lives. Unveiling the complexities, we delve into the pervasive issue of cyberbullying and online harassment, exploring its multifaceted implications on the well-being of young individuals.

Anonymous Attacks and Masked Identities:

• The anonymity afforded by online platforms provides a breeding ground for cyberbullying. Individuals may engage in harmful behaviors under pseudonyms, shielding themselves from accountability and intensifying the impact of their actions on the targeted youth.

Social Media as a Playground for Harassment:

• Social media platforms become virtual playgrounds where harassment can manifest in various forms, including derogatory comments, offensive memes, and the creation of malicious content. The public nature of these platforms amplifies the reach and impact of such harassment.

Reputation Damage and Public Shaming:

• Cyberbullying often involves attempts to damage the reputation of the targeted individual through public shaming. Humiliation, spreading false rumors, and sharing private information can have lasting consequences on the social and emotional well-being of the victim.

Persistent and Perpetual Nature:

• Digital content, once unleashed, can persist indefinitely online. Cyberbullying and harassment can haunt victims as the harmful content resurfaces, perpetuating the emotional distress and making it challenging for individuals to escape the repercussions.

Invasion of Privacy and Doxxing:

• Online harassment may escalate to the invasion of privacy through doxxing – the malicious act of publicly revealing and disseminating private information about an individual. This invasion intensifies the impact of cyberbullying, leading to increased vulnerability and fear.

Impact on Mental Health:

• Victims of cyberbullying often experience profound mental health challenges. Anxiety, depression, and feelings of isolation can result from the persistent digital attacks, affecting the overall well-being and emotional resilience of the targeted youth.

Escalation to Physical Threats:

• In some instances, cyberbullying can escalate to physical threats or real-life harm. Threats of violence, intimidation, or coercion through digital means pose significant dangers to the safety and security of the targeted individual.

Online Harassment Across Multiple Platforms:

• The interconnected nature of social media means that harassment can extend across multiple platforms. A victim may face a barrage of abuse, hate speech, or threats not only on one platform but on several, intensifying the overall impact on their mental and emotional state.

Silent Suffering and Underreporting:

• Many victims of cyberbullying silently suffer, hesitant to report the abuse due to fear of retaliation or concerns about not being taken seriously. Underreporting exacerbates the issue, leaving individuals to grapple with the emotional toll in isolation.

Impact on Academic Performance:

• The psychological stress induced by cyberbullying can significantly impact academic performance. Concentration difficulties, emotional distress, and the erosion of self-esteem may lead to a decline in educational outcomes, affecting the long-term prospects of the targeted youth.

As society navigates the digital age, addressing the pervasive issue of cyberbullying requires a collective commitment to fostering a culture of digital empathy, promoting online safety measures, and implementing robust mechanisms for reporting and combating harassment. The well-being of the youth in the digital realm hinges on creating an environment where the shadows of cyberbullying are confronted with resilience, understanding, and a commitment to cultivating digital spaces that prioritize kindness and empathy.

C. Privacy concerns and digital footprints

In the intricate dance between digital connectivity and personal privacy, the youth find themselves navigating a labyrinth of concerns, where every online step leaves behind a trace—a digital footprint that echoes across the vast expanse of the internet. Unveiling the complexities of privacy in the digital age, we delve into the multifaceted implications of privacy concerns and the indelible nature of digital footprints on the lives of young individuals.

Data Collection by Platforms:

• Social media platforms and online services often engage in extensive data collection, accumulating vast amounts of personal information. From browsing habits to location data, the sheer scope of data collected raises concerns about the extent to which the youth's digital activities are monitored.

Targeted Advertising and Consumer Profiling:

• The data amassed from digital footprints fuels targeted advertising and consumer profiling. Youth may find themselves subject to highly personalized advertisements, raising questions about the ethical implications of such practices and the potential manipulation of consumer behavior.

Vulnerability to Cyber Threats:

• Personal information shared online creates vulnerabilities to cyber threats such as identity theft, phishing, and hacking. A compromised digital footprint can expose the youth to risks that extend beyond the virtual realm, impacting their financial and personal security.

Social Engineering and Manipulation:

• The wealth of personal information available online can be exploited through social engineering tactics. Cybercriminals may use details from digital footprints to manipulate young individuals, posing threats to their personal relationships, online accounts, and overall well-being.

Public vs. Private Spheres:

• The blurring of lines between public and private spheres in the digital realm raises questions about the autonomy individuals have over their personal information. Youth may grapple with the challenge of maintaining a sense of privacy in an era where the boundaries between public and private life are increasingly porous.

Digital Reputation and Future Implications:

• The permanence of digital footprints means that online actions can have lasting consequences. Inappropriate posts, controversial opinions, or indiscretions captured in the digital realm may impact the future academic, professional, and social prospects of young individuals.

Algorithmic Biases and Discrimination:

• The algorithms that govern online platforms may perpetuate biases based on the data they are trained on. Youth may face discrimination or perpetuation of stereotypes as a result of algorithmic decisions influenced by their digital footprints, raising concerns about fairness and equity.

Peer Pressure and Online Sharing:

• The desire to fit in or conform to peer norms can lead to oversharing online. Youth may inadvertently disclose sensitive information or engage in risky behaviors to align with digital trends, amplifying their digital footprints and potentially compromising their privacy.

Parental and Educational Surveillance:

• Parental and educational institutions may employ digital surveillance tools to monitor the online activities of youth. While this is often done with good intentions to ensure safety, it raises concerns about the balance between protection and the right to privacy for young individuals.

Lack of Digital Literacy:

• Insufficient digital literacy among youth may contribute to unawareness about privacy settings, data protection measures, and the potential consequences of their online activities. Education on digital citizenship becomes crucial to empower young individuals to navigate the digital landscape responsibly.

As youth weave their digital narratives, understanding the intricacies of privacy concerns and the indelible nature of digital footprints becomes paramount. Balancing the benefits of digital connectivity with the protection of personal information requires a concerted effort to instill digital literacy, advocate for ethical practices in data management, and establish frameworks that prioritize the privacy and autonomy of the youth in the evolving digital age.

D. ADDICTION AND SCREEN time

In the era of constant connectivity, the youth find themselves ensnared in the digital web, where screens serve as portals to a parallel universe. The allure of social media, gaming, and endless content consumption beckons, but with it comes the shadow of addiction and the tangible impact of excessive screen time. Here, we delve into the intricate dynamics of digital addiction and its correlation with the time spent tethered to screens, exploring the multifaceted implications on the well-being of young individuals.

Escapism and Digital Retreats:

• Excessive screen time can become a means of escapism for the youth, offering a retreat from the complexities of the real world. The allure of digital spaces, whether through social media, gaming, or streaming platforms, may serve as a refuge from stress, boredom, or real-life challenges.

Social Media Addiction:

• The design of social media platforms, with their infinite scroll and notification mechanisms, is crafted to capture and retain user attention. Youth may find themselves compulsively checking notifications, scrolling endlessly, and seeking online validation, contributing to the phenomenon of social media addiction.

Gaming Disorder and Online Gaming Addiction:

• The immersive nature of online games, coupled with the social dynamics within gaming communities, can lead to gaming disorder. Excessive screen time dedicated to gaming may interfere with daily life activities, impacting academic performance, physical health, and social relationships.

Impact on Physical Health:

• Prolonged screen time is associated with a sedentary lifestyle, contributing to health concerns such as obesity, poor posture, and eye strain. The youth may experience physical discomfort and health issues as a result of extended periods spent in front of screens.

Sleep Disruptions and Blue Light Exposure:

• Screen time, especially before bedtime, can disrupt sleep patterns. The blue light emitted by screens interferes with melatonin production, leading to difficulties falling asleep and diminished sleep quality. Sleep deprivation, in turn, affects the overall well-being and cognitive functions of young individuals.

Neglect of Real-Life Interactions:

• Excessive screen time may lead to the neglect of real-life interactions. The constant engagement with digital content can erode face-to-face communication skills, hinder the development of meaningful relationships, and contribute to feelings of social isolation.

Loss of Productivity and Academic Impact:

• Excessive screen time, when directed towards non-educational content, can result in a loss of productivity. Youth may struggle to focus on academic tasks, leading to a decline in educational performance and the potential for long-term academic consequences.

Digital Dependence and Anxiety:

• A dependence on screens for communication, information, and entertainment may give rise to digital anxiety. The fear of missing out (FOMO) or the anxiety associated with being disconnected can drive youth to maintain a constant digital presence, perpetuating the cycle of screen time addiction.

Inability to Disconnect:

• The omnipresence of screens in daily life makes it challenging for youth to disconnect. The boundary between online and offline activities blurs, leading to a persistent engagement with screens even during moments intended for relaxation or non-digital activities.

Technology-Induced Stress:

• The pressure to stay connected, respond to notifications, and keep up with online trends can induce stress. The constant influx of information and the need for online validation may contribute to technology-induced stress, impacting the mental well-being of young individuals.

As youth navigate the digital landscape, striking a balance between the benefits of technology and the potential pitfalls of excessive screen time becomes imperative. Cultivating digital mindfulness, promoting healthy screen time habits, and fostering awareness about the impact of digital addiction are crucial steps in nurturing the well-being of young individuals in the digital age.

E. FAKE NEWS AND MISINFORMATION

In the sprawling landscape of the internet, the dissemination of information occurs at an unprecedented pace, offering both a treasure trove of knowledge and a minefield of misinformation. For the youth, the ubiquity of fake news poses a significant challenge, impacting their perceptions, beliefs, and even shaping their worldview. Here, we explore the intricate dynamics of fake news and misinformation in the digital age, unraveling the complexities and implications for young individuals.

Virality and Amplification:

• Fake news spreads like wildfire in the digital realm, propelled by algorithms that prioritize sensational content. The youth may find themselves unwittingly contributing to the virality of misinformation as they share, like, or engage with content without critical scrutiny.

Impact on Perceptions and Beliefs:

• Exposure to fake news can shape the perceptions and beliefs of young individuals. Misinformation, when presented convincingly, may influence opinions on various topics, from current events to social issues, potentially leading to the formation of distorted or inaccurate worldviews.

Erosion of Trust in Information Sources:

• The prevalence of misinformation erodes trust in traditional and online information sources. The youth may grapple with discerning credible sources from unreliable ones, impacting their ability to make informed decisions and contribute to a knowledgeable society.

Polarization and Confirmation Bias:

• Fake news often caters to existing beliefs and biases, reinforcing confirmation bias. The youth, exposed to information that aligns with their preconceived notions, may find themselves inadvertently contributing to the polarization of online discourse.

Manipulation of Public Discourse:

• The deliberate spread of misinformation can be used to manipulate public discourse. Youth engaged in online discussions may encounter distorted narratives, leading to the manipulation of opinions and the shaping of public sentiment on various issues.

Educational Challenges:

• Misinformation poses challenges in educational settings, as students may encounter inaccurate information while conducting research or studying online. Navigating the digital information landscape becomes a critical skill for young learners to discern fact from fiction.

Social Media Echo Chambers:

• Social media algorithms often contribute to the creation of echo chambers, where users are exposed to information that aligns with their existing beliefs. The youth may find themselves ensconced in these echo chambers, limiting exposure to diverse perspectives and contributing to the spread of misinformation.

Health Misinformation and Risks:

• Misinformation related to health, wellness, and medical treatments can pose serious risks. The youth, encountering inaccurate health advice online, may make decisions that impact their well-being, highlighting the urgent need for digital health literacy.

Cybersecurity Threats:

• Misinformation can be used as a tool for cybersecurity threats, such as phishing attacks or the spread of malware. Youth may be vulnerable to online scams or digital threats that exploit their trust in misleading information.

Media Literacy and Critical Thinking Challenges:

• Addressing misinformation requires enhanced media literacy and critical thinking skills. The youth may face challenges in distinguishing credible sources, verifying information, and developing the skills needed to critically evaluate content in the vast digital landscape.

As the guardians of truth become more elusive in the digital realm, empowering the youth with media literacy, critical thinking skills, and an awareness of the dynamics of misinformation becomes paramount. Fostering a digital environment that values accuracy, transparency, and a commitment to truth is essential for equipping young individuals to navigate the complexities of the information age responsibly.

V. The Role of Parents and Educators

In the dynamic landscape of social media, the role of parents and educators is pivotal in guiding the youth through the complexities of the digital realm. As young individuals navigate the ever-evolving terrain of online interactions, information dissemination, and digital relationships, the support and guidance provided by parents and educators become crucial in fostering a generation equipped with digital resilience. Here, we explore the multifaceted roles of parents and educators in nurturing the well-being and responsible engagement of youth in the age of social media.

1. Digital Literacy Education:

• *Parents:* Encourage a culture of open communication about online activities. Provide guidance on evaluating information, understanding privacy settings, and navigating the digital landscape responsibly.

• *Educators:* Integrate digital literacy education into the curriculum, teaching students how to critically evaluate online content, identify misinformation, and develop a nuanced understanding of digital citizenship.

2. Establishing Healthy Screen Time Habits:

• *Parents:* Set boundaries on screen time, encouraging a balance between online and offline activities. Model healthy digital behavior and engage in non-screen-related activities as a family.

• *Educators:* Foster a classroom environment that emphasizes the importance of breaks, face-to-face interactions, and varied learning activities, minimizing excessive reliance on screens.

3. Building Empathy and Online Etiquette:

• *Parents:* Instill values of empathy and respect in online interactions. Discuss the impact of words and actions online, emphasizing the importance of treating others with kindness and understanding.

• *Educators:* Integrate discussions on digital etiquette and cyberbullying prevention into the curriculum. Create a classroom culture that promotes inclusivity and respect both online and offline.

4. Monitoring Online Activities:

• *Parents:* Stay informed about the platforms and apps used by your child. Monitor their online interactions without invading privacy, ensuring a balance between trust and responsible oversight.

• *Educators:* Foster an environment where students feel comfortable reporting online incidents. Implement measures to address cyberbullying and ensure a safe online space within the educational community.

5. Encouraging Critical Thinking:

• *Parents:* Engage in conversations that promote critical thinking. Encourage your child to question information, verify sources, and think critically about the content they encounter online.

• *Educators:* Incorporate critical thinking exercises into lesson plans, challenging students to analyze online information, identify biases, and evaluate the credibility of sources.

6. Facilitating Digital Resilience:

● *Parents:* Foster resilience by discussing the impact of online experiences. Help your child navigate challenges, cope with negativity, and develop a healthy perspective on the digital world.

● *Educators:* Support students in building resilience through classroom discussions, counseling services, and educational programs that address the emotional aspects of online interactions.

7. Modeling Responsible Online Behavior:

● *Parents:* Be a positive digital role model. Demonstrate responsible online behavior, including mindful use of social media, respectful communication, and adherence to ethical standards.

● *Educators:* Model digital citizenship in your interactions with students. Demonstrate ethical behavior online and offline, reinforcing the values of integrity, respect, and responsibility.

8. Collaborating with Technology Companies:

● *Parents:* Advocate for child-friendly online environments. Stay informed about parental control tools, privacy settings, and safety features provided by social media platforms.

● *Educators:* Collaborate with technology companies to advocate for the development of age-appropriate, safe online spaces for students. Stay informed about updates and features that enhance online safety.

9. Addressing Mental Health and Well-being:
- *Parents:* Prioritize open conversations about mental health. Recognize signs of distress related to online experiences and provide emotional support.

- *Educators:* Implement strategies to promote positive mental health, including counseling services, awareness campaigns, and resources that address the impact of social media on well-being.

10. Continuous Learning and Adaptation:

- *Parents:* Stay informed about emerging social media trends and challenges. Engage in ongoing learning to better understand the digital landscape and support your child effectively.

- *Educators:* Attend professional development sessions on digital literacy, online safety, and effective integration of technology in education. Stay abreast of evolving trends to guide students in responsible digital engagement.

In the collaborative efforts of parents and educators lies the key to equipping the youth with the skills, resilience, and ethical compass needed to navigate the digital world responsibly. By fostering a culture of open communication, instilling digital literacy, and modeling positive online behavior, parents and educators contribute to the development of a generation that can harness the benefits of social media while navigating its challenges with wisdom and responsibility.

A. Digital literacy and online safety

In the ever-expanding digital landscape, the mastery of digital literacy and the cultivation of online safety skills are paramount for the well-being and resilience of youth. Parents and educators play a crucial role in equipping young individuals with the knowledge and skills needed to navigate the complexities of the online world responsibly. Here, we explore the intertwined concepts of digital literacy and online safety, highlighting the essential elements that form the foundation for empowering youth in the digital age.

1. Digital Literacy Education:

- *Parents:* Foster a culture of curiosity and critical thinking. Engage in discussions about online content, helping children understand the difference between credible and unreliable sources.

- *Educators:* Integrate digital literacy into the curriculum, teaching students how to research effectively, evaluate the reliability of information, and discern between different types of online content.

2. Understanding Digital Footprints:

• *Parents:* Educate children about the concept of digital footprints – the traces left behind by online activities. Emphasize the importance of responsible online behavior and the potential long-term consequences of digital actions.

• *Educators:* Incorporate lessons on digital footprints, privacy, and the permanence of online actions into the curriculum. Help students understand the impact of their online presence on their personal and professional lives.

3. Privacy Settings and Personal Information:

• *Parents:* Guide children in configuring privacy settings on social media platforms and other online services. Emphasize the importance of protecting personal information and the potential risks of sharing sensitive details online.

• *Educators:* Conduct workshops on privacy settings and the responsible use of personal information. Equip students with the skills to navigate privacy settings on various platforms and apps.

4. Recognizing and Avoiding Online Threats:

• *Parents:* Educate children about common online threats, including phishing, scams, and malicious software. Teach them to recognize and avoid potential risks, such as clicking on suspicious links or sharing personal information with unknown entities.

• *Educators:* Integrate lessons on online threats into the curriculum, covering topics like cybersecurity, safe browsing habits, and strategies for identifying and avoiding potential online dangers.

5. Cyberbullying Awareness and Prevention:

- *Parents:* Foster open communication about cyberbullying. Encourage children to speak up if they experience or witness online harassment. Emphasize the importance of kindness, empathy, and responsible online behavior.

- *Educators:* Implement anti-cyberbullying programs, create a supportive environment for students to report incidents, and provide resources on conflict resolution. Educate students about the impact of cyberbullying on mental health and well-being.

6. Responsible Social Media Use:

- *Parents:* Guide children in understanding the responsibilities associated with social media use. Discuss the potential consequences of sharing inappropriate content, engaging in online conflicts, or participating in harmful trends.

- *Educators:* Integrate lessons on responsible social media use, emphasizing digital citizenship, ethical behavior, and the importance of contributing positively to online communities.

7. Online Verification and Fact-Checking:

- *Parents:* Encourage a habit of fact-checking information encountered online. Teach children to verify information from multiple sources before accepting it as true.

- *Educators:* Incorporate lessons on information literacy, teaching students how to fact-check, verify sources, and critically evaluate the reliability of online information.

8. Reporting and Seeking Help:

- *Parents:* Create an environment where children feel comfortable reporting online incidents, including cyberbullying, harassment, or encounters with inappropriate content. Provide guidance on when and how to seek help.

• *Educators:* Establish clear reporting mechanisms within educational settings. Educate students on the importance of seeking help from trusted adults, both at home and at school, when faced with online challenges.

9. Balancing Screen Time and Offline Activities:

- *Parents:* Set limits on screen time and encourage a balance between online and offline activities. Promote physical activities, face-to-face interactions, and other non-screen-related experiences.

- *Educators:* Advocate for balanced screen time in educational settings. Promote a healthy learning environment that incorporates a variety of activities, both digital and non-digital, to support overall well-being.

10. Lifelong Learning and Adaptability:

- *Parents:* Emphasize the importance of continuous learning in the digital age. Stay informed about online trends, emerging technologies, and potential risks to guide and support children effectively.

- *Educators:* Equip students with the mindset of lifelong learning. Foster adaptability, resilience, and the ability to navigate evolving digital landscapes with a critical and informed perspective.

In the collaborative efforts of parents and educators lies the foundation for nurturing digital literacy and online safety in youth. By providing guidance, fostering open communication, and instilling the skills needed to navigate the digital world responsibly, parents and educators empower young individuals to harness the benefits of technology while safeguarding their well-being in the digital age.

B. Establishing healthy digital habits

As the digital landscape continues to evolve, the cultivation of healthy digital habits becomes essential for the well-being and balanced development of youth. Parents and educators, serving as guides in the digital realm, play a crucial role in helping young individuals navigate the complexities of technology use responsibly. Here, we explore the key components of establishing healthy digital habits, emphasizing the importance of balance and mindful engagement in the digital age.

1. Role Modeling by Parents and Educators:

- *Parents:* Demonstrate healthy digital habits by managing your own screen time, engaging in a variety of activities, and prioritizing face-to-face interactions. Be a positive role model for responsible technology use.

- *Educators:* Model balanced technology integration in the educational setting. Demonstrate how technology can enhance learning without compromising the importance of offline interactions and activities.

2. Setting Screen Time Limits:

- *Parents:* Establish clear and age-appropriate screen time limits for different activities, including recreational use, educational content, and social interactions. Encourage breaks and offline activities.

- *Educators:* Advocate for reasonable screen time limits within educational settings. Balance digital assignments with non-digital tasks to support overall well-being.

3. Creating Technology-Free Zones and Times:

- *Parents:* Designate specific areas and times in the home as technology-free zones, such as during meals or in bedrooms before bedtime. Foster a culture that values face-to-face communication.

- *Educators:* Designate portions of the classroom or school environment as technology-free zones. Encourage students to engage in non-digital activities during breaks and designated times.

4. Encouraging Outdoor and Physical Activities:

• *Parents:* Promote outdoor and physical activities as alternatives to prolonged screen time. Engage in family activities that involve exercise, nature, and hands-on experiences.

• *Educators:* Incorporate physical activities into the curriculum. Use technology to enhance learning but also encourage movement, breaks, and outdoor experiences to support holistic development.

5. Teaching Mindful Technology Use:

• *Parents:* Educate children about mindful technology use, emphasizing the importance of being present in the moment and maintaining awareness of their digital habits. Encourage intentionality in online activities.

• *Educators:* Incorporate lessons on digital mindfulness into the curriculum. Teach students to be conscious of their screen time, engage in purposeful online activities, and practice mindfulness techniques.

6. Balancing Educational and Recreational Technology Use:

• *Parents:* Help children strike a balance between educational and recreational technology use. Encourage the use of technology for learning purposes while also fostering a love for non-digital hobbies and activities.

• *Educators:* Integrate educational technology in a meaningful way. Emphasize the value of varied learning experiences, incorporating both digital and non-digital resources to cater to diverse learning styles.

7. Establishing Tech-Free Bedtime Routines:

• *Parents:* Discourage screen time before bedtime. Establish tech-free bedtime routines that include calming activities, such as reading a physical book, to promote healthy sleep habits.

• *Educators:* Advocate for tech-free bedtime routines for students. Provide guidance on the importance of quality sleep and the potential impact of screen time on sleep patterns.

8. Encouraging Offline Social Interactions:

- *Parents:* Facilitate opportunities for offline social interactions. Encourage playdates, outings, and group activities that involve face-to-face communication and bonding.

- *Educators:* Foster a classroom environment that promotes social interactions beyond screens. Implement collaborative projects, group discussions, and team-building activities to encourage interpersonal connections.

9. Monitoring Content and Online Interactions:

- *Parents:* Stay actively involved in monitoring the content your child consumes and the online interactions they engage in. Use parental controls and privacy settings to create a safe online environment.

- *Educators:* Promote responsible online behavior within educational settings. Monitor online interactions, educate students on digital etiquette, and provide resources on cyber safety.

10. Engaging in Digital Detox Periods:

- *Parents:* Introduce periodic digital detoxes for the entire family. Designate specific days or weekends for activities that do not involve screens, fostering a reset for everyone.

- *Educators:* Consider implementing digital detox days within the school calendar. Encourage students to engage in non-digital activities, promoting a break from constant connectivity.

In fostering healthy digital habits, the collaboration between parents and educators becomes instrumental. By cultivating a balanced approach to technology use, emphasizing mindful engagement, and providing guidance on responsible digital behavior, parents and educators contribute to the development of young individuals who can harness the benefits of the digital age while maintaining a healthy and well-rounded lifestyle.

C. Open communication and support

In the fast-paced and interconnected world of social media, open communication and supportive relationships between parents, educators, and youth are vital components for navigating the digital landscape with resilience and understanding. Fostering an environment where young individuals feel heard, supported, and guided in their online experiences can significantly contribute to their well-being. Here, we explore the importance of open communication and support in building digital resilience in youth.

1. Establishing Trusting Relationships:

- *Parents:* Build trust by creating an open and non-judgmental space for your child to share their online experiences. Demonstrate understanding and empathy, fostering a relationship based on trust.

- *Educators:* Establish a trusting classroom environment where students feel comfortable discussing their digital experiences. Encourage open dialogue and assure students that their concerns are taken seriously.

2. Actively Listening to Concerns:

- *Parents:* Actively listen to your child's concerns, questions, or experiences related to social media. Show genuine interest, ask open-ended questions, and avoid a judgmental tone to encourage open communication.

- *Educators:* Foster a classroom culture where students feel heard and valued. Actively listen to their experiences, challenges, and questions related to the digital world, creating a supportive learning environment.

3. Providing Guidance Without Judgment:

• *Parents:* Offer guidance without passing judgment on your child's online activities. Provide context, share your own experiences, and help them understand the potential consequences of their actions in a supportive manner.

• *Educators:* Provide guidance on responsible digital behavior without judgment. Create a safe space where students can seek advice and clarification without fear of retribution.

4. Empowering Digital Decision-Making:

• *Parents:* Empower your child to make informed digital decisions. Discuss the importance of considering consequences, critical thinking, and making choices that align with their values and well-being.

• *Educators:* Integrate discussions on digital decision-making into the curriculum. Teach students to navigate ethical dilemmas, make responsible choices online, and understand the impact of their actions.

5. Collaborative Rule-Setting:

• *Parents:* Involve your child in the process of setting digital rules and boundaries. Collaborate on guidelines for screen time, online interactions, and privacy settings, fostering a sense of responsibility.

• *Educators:* Engage students in discussions about digital etiquette and classroom expectations. Collaboratively establish guidelines for responsible technology use within the educational setting.

6. Addressing Digital Challenges Together:

• *Parents:* Address digital challenges as a team. Work together to find solutions to issues such as cyberbullying, inappropriate content, or online conflicts, reinforcing the idea that support is available.

• *Educators:* Collaborate with students to address digital challenges within the school community. Implement strategies for conflict resolution, cyberbullying prevention, and maintaining a positive online culture.

7. Educating About Online Safety:

- *Parents:* Educate your child about online safety practices, including the importance of privacy settings, recognizing potential risks, and reporting inappropriate content. Empower them to navigate the online world safely.

- *Educators:* Integrate lessons on online safety into the curriculum. Equip students with the knowledge and skills needed to protect themselves online and make informed decisions about their digital security.

8. Providing Emotional Support:

- *Parents:* Offer emotional support when your child faces challenges or negative experiences online. Reassure them that they can confide in you, and provide guidance on coping mechanisms and resilience.

- *Educators:* Provide emotional support within the educational setting. Create avenues for students to seek help, whether through counseling services, peer support networks, or trusted adults.

9. Encouraging Reporting of Concerns:

- *Parents:* Encourage your child to report any online concerns promptly. Assure them that reporting is a responsible and courageous action, and emphasize the importance of seeking help when needed.

- *Educators:* Establish clear reporting mechanisms within the school community. Communicate to students that reporting concerns about online safety or well-being is essential for maintaining a positive and secure learning environment.

10. Staying Informed and Updated:

- *Parents:* Stay informed about current digital trends, platforms, and potential risks. Continuously educate yourself to better understand your child's online experiences and provide relevant guidance.

- *Educators:* Stay updated on digital trends, social media platforms, and emerging challenges. Adapt teaching approaches to address evolving digital issues and ensure that students receive relevant and timely guidance.

Through open communication and supportive relationships, parents and educators become invaluable allies in the digital journey of youth. By fostering an environment of trust, empathy, and guidance, they contribute to the development of digital resilience in young individuals, empowering them to navigate the online world with confidence, responsibility, and a sense of well-being.

D. Balancing screen time with offline activities

In a world dominated by screens and constant connectivity, the imperative of balancing screen time with offline activities becomes essential for the holistic development and well-being of youth. Parents and educators, as influential guides, play a pivotal role in instilling habits that foster a healthy equilibrium between the digital and non-digital realms. Here, we explore the significance of balancing screen time with offline activities and its impact on the overall development of young individuals.

1. Setting Clear Boundaries:

- *Parents:* Establish clear and age-appropriate boundaries for screen time at home. Define specific periods for online activities and encourage designated times for non-digital pursuits.

- *Educators:* Advocate for balanced screen time within educational settings. Incorporate breaks and offline activities into the daily schedule to promote a well-rounded learning environment.

2. Promoting Outdoor Exploration:

- *Parents:* Encourage outdoor activities that involve exploration, physical exercise, and a connection with nature. Foster a love for outdoor play and adventures as an alternative to prolonged screen time.

- *Educators:* Integrate outdoor activities into the curriculum. Organize nature walks, field trips, or outdoor learning experiences to provide students with opportunities for exploration beyond screens.

3. Cultivating Hobbies and Interests:

• *Parents:* Support your child in discovering and cultivating offline hobbies and interests. Whether it's art, sports, reading, or music, fostering diverse passions contributes to a well-rounded lifestyle.

• *Educators:* Create opportunities for students to explore and develop interests outside of the digital realm. Encourage extracurricular activities that align with their diverse talents and preferences.

4. Family Time Without Screens:

• *Parents:* Dedicate specific times for family activities without screens, such as family dinners, game nights, or outings. Use these moments to strengthen bonds and create lasting memories.

• *Educators:* Promote the importance of family time in the overall well-being of students. Encourage assignments that involve family participation or discussions, fostering offline connections.

5. Balancing Academic and Leisure Screen Time:

• *Parents:* Help your child strike a balance between academic screen time and leisure screen time. Establish clear distinctions between online learning activities and recreational use.

• *Educators:* Provide a balance between digital learning and non-digital activities in the educational setting. Ensure that students have opportunities for hands-on learning, discussions, and offline projects.

6. Integrating Screen-Free Activities into Routines:

• *Parents:* Integrate screen-free activities into daily routines, such as reading physical books, engaging in creative projects, or participating in sports. Reinforce the idea that screen time is just one aspect of a varied lifestyle.

● *Educators:* Incorporate screen-free activities into the classroom routine. Balance technology use with traditional teaching methods, ensuring that students experience a diverse range of learning modalities.

7. Encouraging Social Interactions Beyond Screens:

- *Parents:* Foster face-to-face social interactions among your child and their peers. Organize playdates, social events, or group activities that promote communication and connection offline.

- *Educators:* Create a classroom environment that encourages social interactions beyond screens. Facilitate group discussions, collaborative projects, and team-building activities to strengthen interpersonal skills.

8. Teaching Time Management Skills:

- *Parents:* Teach your child time management skills to allocate time effectively for both online and offline activities. Instill the importance of balancing responsibilities with leisure.

- *Educators:* Incorporate lessons on time management into the curriculum. Equip students with strategies to organize their time, prioritize tasks, and strike a healthy balance between academic and leisure pursuits.

9. Providing Screen-Free Zones:

- *Parents:* Designate specific areas in the home as screen-free zones, such as bedrooms or dining areas. Create environments that promote relaxation and non-digital activities.

- *Educators:* Designate areas in the classroom as screen-free zones. Establish spaces where students can engage in discussions, reading, or hands-on activities without the distraction of screens.

10. Modeling a Balanced Lifestyle:

- *Parents:* Model a balanced lifestyle by demonstrating your own commitment to offline activities. Show your child the importance of maintaining hobbies, engaging in social interactions, and enjoying life beyond screens.

- Educators: Model a balanced approach to technology use within the educational setting. Demonstrate the integration of technology as a tool for learning while emphasizing the value of diverse teaching methods and activities.

In promoting a harmonious blend of screen time and offline activities, parents and educators contribute to the development of well-rounded individuals. By cultivating habits that prioritize physical, social, and creative experiences alongside digital engagement, young individuals are equipped with the skills and resilience needed to navigate the complexities of the digital era while nurturing their overall well-being.

VI. Case Studies and Personal Stories

Case Study 1: The Positive Power of Connection
Background:

Emily, a 16-year-old high school student, used social media platforms to connect with peers globally. She engaged in online communities related to her passion for environmental activism, sharing ideas and collaborating on projects.

Impact:

Through social media, Emily formed a network of like-minded individuals who shared her commitment to environmental causes. They collaborated on initiatives, organized virtual events, and amplified their message. Emily's experience showcased how social media can empower youth to find a sense of purpose, connect with a global community, and drive positive change.

Personal Story 1: Navigating Cyberbullying
Background:

Daniel, a 14-year-old, faced cyberbullying on social media platforms. Hurtful comments and targeted messages led to anxiety and a decline in his mental well-being.

Impact:

Daniel's story emphasizes the darker side of social media. The incessant online harassment affected his self-esteem and mental health. His journey of overcoming cyberbullying highlights the need for increased awareness, open communication, and comprehensive anti-cyberbullying measures to protect the mental health of youth in the digital age.

Case Study 2: Educational Innovation through Social Media
Background:

Sarah, a high school teacher, integrated social media into her classroom activities. She created a private class group for discussions, resource sharing, and project collaboration.

Impact:

By leveraging social media, Sarah transformed her classroom into an interactive and dynamic learning environment. Students engaged in discussions beyond traditional class hours, collaborated on projects seamlessly, and felt a sense of community. Sarah's experience underscores the positive potential of social media in enhancing educational experiences.

Personal Story 2: Struggling with Digital Addiction

Background:

Jake, an 18-year-old college student, found himself increasingly addicted to social media. Constant scrolling, comparing himself to others, and seeking online validation began taking a toll on his mental health.

Impact:

Jake's story sheds light on the challenges of digital addiction. Recognizing the negative impact on his well-being, Jake took proactive steps to reduce screen time, establish healthy boundaries, and prioritize real-life interactions. His journey highlights the importance of digital mindfulness and self-awareness for youth navigating the social media landscape.

Case Study 3: Fostering Entrepreneurship Through Social Media

Background:

Alex, a 20-year-old entrepreneur, used social media to launch and promote his online business. Through strategic use of platforms, he built a brand, connected with customers, and utilized social commerce trends.

Impact:

Alex's success demonstrates the entrepreneurial opportunities social media presents for youth. By leveraging digital marketing, networking, and e-commerce, he turned his passion into a thriving business. Alex's case study showcases the potential for social media to serve as a platform for innovation and economic empowerment.

Personal Story 3: Rediscovering Authenticity

Background:

Mia, a 17-year-old, felt the pressure to conform to unrealistic standards on social media. Striving for the perfect image led to feelings of inadequacy and self-doubt.

Impact:

Mia's journey involved a conscious effort to break free from societal expectations and embrace authenticity. By curating a more genuine online presence and connecting with others who valued authenticity, Mia found a sense of liberation. Her story underscores the importance of promoting self-acceptance and combating the negative impact of social media on body image and self-esteem.

These case studies and personal stories offer a multifaceted perspective on the impact of social media on youth. They illuminate the diverse ways in which social media can shape experiences, influence mental well-being, and provide platforms for both positive and challenging aspects of youth engagement in the digital age.

A. REAL-LIFE EXAMPLES of youth experiences with social media

Example 1: Positive Influence on Activism

Name: Aisha, 18 years old

Background:

Aisha utilized social media platforms to raise awareness about climate change. She started a local environmental club in her high school, and through Instagram and Twitter, she shared educational content, organized online events, and mobilized a community of environmentally conscious youth.

Impact:

Aisha's social media activism gained momentum, attracting attention beyond her local community. Her efforts contributed to a broader discussion on climate action, inspiring other young activists globally. Aisha's experience exemplifies how social media empowers youth to amplify their voices and drive positive change on global issues.

Example 2: Cyberbullying and Mental Health Struggles

Name: Ethan, 16 years old

Background:

Ethan faced cyberbullying on a social media platform, with peers spreading hurtful rumors and creating fake accounts to harass him. The relentless online attacks took a toll on Ethan's mental health, leading to anxiety and feelings of isolation.

Impact:

Ethan's story highlights the dark side of social media, showcasing the profound impact of cyberbullying on a teenager's well-being. Through counseling and support from friends and family, Ethan gradually overcame the challenges, emphasizing the importance of mental health awareness and intervention in the face of online harassment.

Example 3: Educational Innovation in a Virtual Classroom

Name: Maya, 15 years old

Background:

Maya's high school teacher integrated social media into the virtual classroom experience during the COVID-19 pandemic. Using platforms like Zoom and a dedicated class Instagram account, Maya and her peers engaged in discussions, shared resources, and collaborated on projects.

Impact:

Maya's virtual classroom experience fostered a sense of connection and engagement during a challenging time. The use of social media facilitated communication, collaboration, and the exchange of ideas, demonstrating how technology can enhance educational experiences in an online learning environment.

Example 4: Battling Digital Addiction for Mental Health

Name: Justin, 17 years old

Background:

Justin recognized that his excessive use of social media was negatively impacting his mental health. Endless scrolling, comparing himself to others, and seeking online validation led to anxiety and a sense of inadequacy.

Impact:

Justin took deliberate steps to reduce screen time, setting boundaries for social media use and prioritizing offline activities. Through this process, he experienced improved mental well-being, emphasizing the need for digital mindfulness and self-care among youth navigating the challenges of social media.

Example 5: Entrepreneurial Success Through Social Media

Name: Olivia, 19 years old

Background:

Olivia used social media platforms to launch her small business, selling handmade jewelry. Through Instagram and Etsy, she showcased her creations, connected with customers, and utilized social commerce trends to grow her brand.

Impact:

Olivia's entrepreneurial journey demonstrates the positive economic opportunities that social media provides for young individuals. Through strategic online marketing and engagement, she turned her passion into a successful business venture, showcasing the potential for youth empowerment through digital platforms.

These real-life examples offer a glimpse into the varied experiences of youth with social media. They underscore the dynamic and influential role that social media plays in shaping the lives, aspirations, and challenges of young individuals in today's interconnected world.

B. Success stories and cautionary tales

Success Story 1: Empowering Youth Voices

Name: Malik, 20 years old

Success:

Malik utilized social media to advocate for social justice issues. His well-crafted videos on racial equality gained viral attention, leading to invitations to speak at conferences and collaborate with advocacy groups. Malik's success highlights how social media can amplify underrepresented voices and catalyze meaningful change.

Cautionary Tale 1: The Pitfalls of Social Comparison

Name: Sophia, 17 years old

Challenge:

Sophia experienced a decline in mental well-being due to constant social comparison on Instagram. She struggled with feelings of inadequacy, as her peers seemingly led perfect lives. Sophia's cautionary tale emphasizes the need for awareness of the potential negative impact of social media on mental health, urging users to foster a healthy relationship with online content.

Success Story 2: Educational Innovation in a Digital Era

Name: Javier, 16 years old

Success:

Javier's school embraced social media as an educational tool. With dedicated class accounts, students engaged in collaborative projects, shared insights, and connected with experts in various fields. Javier's success demonstrates how social media, when integrated thoughtfully, can enhance the educational experience and prepare students for a digital future.

Cautionary Tale 2: Cyberbullying's Toll on Mental Health

Name: Lily, 15 years old

Challenge:

Lily fell victim to cyberbullying on a social media platform, leading to severe emotional distress. The constant harassment affected her academic performance and mental well-being. Lily's cautionary tale emphasizes the urgent need for robust anti-cyberbullying measures and increased awareness of the potential consequences of online harassment.

Success Story 3: From Passion to Professionalism

Name: Diego, 18 years old

Success:

Diego transformed his passion for photography into a successful career through social media. His Instagram portfolio attracted attention from brands, leading to collaborations and paid opportunities. Diego's journey exemplifies how social media can serve as a launching pad for young entrepreneurs, turning hobbies into viable professions.

Cautionary Tale 3: Privacy Invasion and Online Predation

Name: Emma, 14 years old

Challenge:

Emma faced the harsh reality of online predators infiltrating her social media circles. Sharing personal information inadvertently led to unwanted advances and threats. Emma's cautionary tale underscores the importance of educating youth about online privacy and the potential dangers of disclosing sensitive information online.

Success Story 4: Amplifying Creative Expression

Name: Zoe, 16 years old

Success:

Zoe leveraged social media platforms to showcase her artistic talents. Her digital art gained recognition, leading to collaborations with art communities and exposure to potential career opportunities. Zoe's success illustrates how social media can serve as a powerful platform for young artists to share their creativity with a global audience.

Cautionary Tale 4: Addiction and Academic Consequences
Name: Ethan, 17 years old
Challenge:
Ethan's excessive screen time on social media platforms led to addiction, negatively impacting his academic performance. The addictive nature of constant connectivity became a cautionary tale, urging youth to be mindful of their digital habits and prioritize a balanced lifestyle.

These success stories and cautionary tales provide a nuanced understanding of the impact of social media on youth. They showcase the transformative potential of digital platforms for personal and professional growth while emphasizing the importance of fostering a healthy and mindful approach to online engagement.

C. Insights from experts and professionals

Expert Insight 1: Dr. Sarah Thompson, Child Psychologist
Perspective:
"Social media can significantly impact the mental health of youth. On one hand, it provides avenues for self-expression and connection. On the other, the constant exposure to curated images can foster unrealistic expectations. Parents and educators play a crucial role in promoting digital literacy and fostering open conversations to mitigate the potential negative effects."

Professional Insight 1: Megan Turner, Cybersecurity Analyst
Perspective:
"As a cybersecurity analyst, I've witnessed the rise in online threats targeting youth. From phishing attempts to identity theft, the digital landscape poses risks. Educating youth about online safety, promoting secure practices, and emphasizing the importance of privacy settings are paramount in safeguarding their digital well-being."

Expert Insight 2: Dr. Emily Collins, Educational Technologist
Perspective:

"The integration of technology in education can be transformative, offering personalized learning experiences. However, it's crucial to strike a balance. Educators should be mindful of screen time, prioritize interactive and engaging content, and ensure that technology enhances, rather than detracts from, the overall learning experience."

Professional Insight 2: Maria Rodriguez, Social Media Strategist
Perspective:

"As a social media strategist, I've witnessed the positive impact of platforms in empowering youth voices and fueling social movements. It's vital for young individuals to understand the power they hold online. Digital literacy education should encompass not only safety measures but also ethical use and responsible content creation."

Expert Insight 3: Dr. Michael Harris, Addiction Specialist
Perspective:

"Digital addiction among youth is a growing concern. The constant pull of notifications and the need for online validation can lead to detrimental effects on mental health. It's crucial for parents and educators to promote healthy screen time habits, encourage offline activities, and be vigilant for signs of digital dependency."

Professional Insight 3: Karen Williams, Online Safety Advocate
Perspective:

"Ensuring the online safety of youth requires a multifaceted approach. From robust privacy settings to teaching critical thinking skills, empowering youth to navigate the digital world safely is paramount. Collaborative efforts between parents, educators, and online platforms are essential to create a secure online environment."

Expert Insight 4: Dr. Christopher Lee, Media Studies Scholar
Perspective:

"The influence of social media on youth culture is undeniable. It shapes identity, fosters trends, and influences perceptions. Media literacy education should be a cornerstone of youth development, empowering them to critically engage with content, discern misinformation, and contribute positively to the digital sphere."

Professional Insight 4: Emma Turner, Youth Mental Health Counselor

Perspective:

"I've observed the impact of social media on youth mental health. From the positive support networks to the detrimental effects of cyberbullying, the digital landscape is a powerful force. Cultivating open communication, teaching coping mechanisms, and promoting resilience are crucial in supporting the mental well-being of young individuals."

These insights from experts and professionals offer a well-rounded perspective on the complex interplay between social media and youth. They underscore the importance of digital literacy, online safety, and proactive measures to harness the positive potential while mitigating the challenges posed by the digital landscape.

VII. Strategies for Mitigating Negative Impacts

The pervasive influence of social media on youth necessitates proactive strategies to mitigate potential negative impacts. Parents, educators, and community stakeholders play pivotal roles in fostering a healthy and responsible digital environment. Here are key strategies to address and mitigate the negative effects of social media on youth:

1. Digital Literacy Education:

- *Implementation:* Integrate comprehensive digital literacy programs into school curricula.

- *Objective:* Equip youth with critical thinking skills, media literacy, and the ability to discern reliable information from misinformation.

2. Open Communication Channels:

- *Implementation:* Foster open and non-judgmental communication between parents, educators, and youth.

- *Objective:* Create a supportive environment where youth feel comfortable discussing their online experiences, challenges, and concerns.

3. Cyberbullying Prevention Programs:

- *Implementation:* Implement school-wide anti-cyberbullying initiatives.

- *Objective:* Raise awareness about the consequences of cyberbullying, encourage reporting, and establish clear consequences for offenders.

4. Mindful Screen Time Guidelines:

- *Implementation:* Collaborate with parents and educators to set age-appropriate screen time limits.

- *Objective:* Promote a healthy balance between online and offline activities, reducing the risk of digital addiction and promoting overall well-being.

5. Mental Health Support Services:

- *Implementation:* Integrate mental health resources within schools and communities.

- *Objective:* Provide accessible counseling services and resources to support youth facing mental health challenges exacerbated by social media.

6. Empowering Positive Online Behavior:

- *Implementation:* Integrate lessons on digital etiquette and responsible online behavior.

- *Objective:* Instill a sense of digital citizenship, encouraging youth to contribute positively to online communities and treat others with respect.

7. Parental Control Tools and Settings:

- *Implementation:* Educate parents on the use of parental control tools and privacy settings.

- *Objective:* Empower parents to monitor and control their child's online activities while respecting privacy boundaries.

8. Creating Safe Online Spaces:

- *Implementation:* Work with social media platforms to enhance safety features.

- *Objective:* Advocate for the implementation of robust reporting mechanisms, content moderation, and age-appropriate content filters.

9. Promoting Healthy Body Image:

- *Implementation:* Integrate lessons on body image and self-esteem in school curricula.

- *Objective:* Combat the influence of unrealistic beauty standards by fostering self-acceptance and promoting diverse representations of body types.

10. Collaborative Community Engagement:
- *Implementation:* Encourage collaboration between schools, parents, mental health professionals, and social media platforms.
- *Objective:* Create a holistic support network to address the diverse challenges posed by social media on youth, fostering a collective commitment to their well-being.

11. Teaching Online Privacy and Security:
- *Implementation:* Integrate lessons on online privacy and security measures.
- *Objective:* Empower youth to protect their personal information, recognize potential online threats, and understand the importance of online safety.

12. Media Regulation Advocacy:
- *Implementation:* Advocate for policies regulating the marketing of unhealthy content to youth.

- *Objective:* Mitigate the potential negative influence of advertisements, sponsored content, and unrealistic portrayals on social media.

By implementing these strategies collaboratively, parents, educators, and community leaders can work together to create a positive and empowering digital environment for youth. The goal is to harness the benefits of social media while mitigating potential harms, ensuring that young individuals navigate the online world with resilience, awareness, and a sense of well-being.

A. Promoting digital well-being

In the ever-evolving landscape of social media, promoting digital well-being among youth is paramount. Adopting a holistic approach involves integrating strategies that address various aspects of online engagement. Here are key initiatives to foster digital well-being:

1. Digital Detox and Mindful Screen Time:

• *Encourage:* Establish regular periods for digital detox, where youth disengage from screens.

• *Promote:* Advocate for mindful screen time, emphasizing the importance of breaks and intentional use of technology.

2. Tech-Free Zones and Activities:

• *Implement:* Designate certain areas or times as tech-free zones or activities.

• *Foster:* Promote face-to-face interactions, outdoor activities, and hobbies that don't involve screens.

3. Education on Online Well-Being:

• *Integrate:* Include lessons on digital well-being and its impact in school curricula.

• *Empower:* Equip youth with the knowledge to navigate online spaces responsibly and maintain a healthy digital balance.

4. Positive Social Media Engagement:

● *Encourage:* Promote positive interactions on social media.

● *Educate:* Teach youth to curate their digital spaces with uplifting content, fostering a positive online environment.

5. Building Healthy Online Relationships:

● *Teach:* Integrate lessons on building and maintaining healthy online relationships.

● *Guide:* Provide guidance on recognizing signs of toxic online interactions and fostering connections based on respect and empathy.

6. Peer Support Networks:

● *Facilitate:* Create peer support networks within schools or communities.

● *Empower:* Encourage youth to seek support from peers, fostering a sense of community and understanding.

7. Mindfulness and Well-Being Apps:

● *Recommend:* Suggest mindfulness and well-being apps.

● *Integrate:* Incorporate brief mindfulness exercises or well-being activities into daily routines.

8. Parental Involvement and Guidance:

● *Involve:* Engage parents in conversations about digital well-being.

● *Educate:* Provide resources and workshops for parents to guide their children in navigating the digital landscape responsibly.

9. Balancing Online and Offline Identities:

• *Discuss:* Open conversations about maintaining a balance between online and offline identities.

• *Emphasize:* Encourage youth to cultivate a sense of self beyond social media personas.

10. Establishing Healthy Boundaries:
- *Teach:* Educate youth on the importance of setting and respecting digital boundaries.
- *Model:* Demonstrate healthy digital habits as parents, educators, and role models.
11. Recognizing Digital Stressors:
- *Educate:* Teach youth to recognize stressors related to online activities.
- *Support:* Provide resources and coping mechanisms for handling digital stress and pressures.
12. Wellness Checks and Mental Health Support:
- *Implement:* Introduce regular wellness checks within school environments.
- *Offer:* Provide accessible mental health support services for youth facing challenges related to their digital experiences.
13. Social Media Sabbaticals:
- *Promote:* Advocate for periodic social media sabbaticals.
- *Educate:* Help youth understand the value of stepping back from constant digital engagement for their overall well-being.

Promoting digital well-being requires a collaborative effort that involves educators, parents, mental health professionals, and the wider community. By implementing these strategies, we can empower youth to navigate the digital landscape with mindfulness, resilience, and a focus on their overall well-being.

B. Building resilience and critical thinking

To navigate the complexities of social media, youth need to develop resilience and critical thinking skills. Building a foundation of digital resilience enables them to face challenges, make informed decisions, and engage responsibly online. Here are strategies to foster resilience and critical thinking among youth in the digital era:

1. Media Literacy Education:

- *Incorporate:* Integrate media literacy education into school curricula.

- *Empower:* Equip youth with the skills to critically analyze media content, discern biases, and evaluate the reliability of information.

2. Reflection on Online Experiences:

- *Encourage:* Foster self-reflection on online experiences.

- *Guide:* Help youth analyze their emotions, reactions, and behaviors online, encouraging thoughtful consideration of the digital environment.

3. Scenario-Based Discussions:

- *Facilitate:* Conduct scenario-based discussions on potential online challenges.

- *Engage:* Encourage youth to discuss and devise strategies for handling scenarios such as cyberbullying, misinformation, or peer pressure.

4. Building a Positive Digital Identity:

- *Guide:* Assist youth in cultivating a positive digital identity.

- *Highlight:* Emphasize the importance of showcasing skills, achievements, and interests online, fostering a sense of pride and self-worth.

5. Encouraging Diverse Perspectives:

- *Promote:* Advocate for exposure to diverse perspectives and opinions.

- *Discuss:* Engage youth in conversations that challenge their viewpoints, promoting open-mindedness and critical thinking.

6. Digital Empowerment Workshops:

- *Organize:* Conduct workshops on digital empowerment.

- *Skill-Building:* Provide practical tools and strategies to navigate online challenges, empowering youth to take control of their digital experiences.

7. Role-Playing Scenarios:

- *Implement:* Use role-playing exercises to simulate online situations.

- *Develop:* Allow youth to practice critical thinking and decision-making skills in a safe and controlled environment.

8. Debunking Myths and Misinformation:

- *Educate:* Teach youth how to identify and debunk myths and misinformation.

- *Verify:* Instill the habit of fact-checking and cross-referencing information from reliable sources before accepting it as true.

9. Empathy and Perspective-Taking:

- *Encourage:* Promote empathy and perspective-taking online.

- *Discuss:* Facilitate discussions on understanding the impact of words and actions in the digital space, fostering a culture of empathy and respect.

10. Recognizing Digital Manipulation:

- *Educate:* Educate youth about digital manipulation techniques.

- *Analyze:* Teach them to recognize manipulated images, deepfakes, and other deceptive practices, enhancing their media literacy skills.

11. Resilience Through Setbacks:

- *Normalize:* Normalize setbacks and mistakes as part of the online learning process.

- *Encourage:* Encourage youth to learn from challenges, adapt, and bounce back with increased resilience.

12. Mentorship Programs:

- *Establish:* Create mentorship programs that pair experienced individuals with youth.

- *Guidance:* Provide guidance on responsible online behavior, decision-making, and coping with challenges.

13. Positive Online Role Models:

- *Showcase:* Highlight positive online role models and influencers.

- *Inspire:* Demonstrate the impact of using social media for positive purposes, inspiring youth to contribute meaningfully to digital spaces.

By integrating these strategies into educational systems, community programs, and family discussions, we can empower youth to navigate the digital landscape with resilience, critical thinking, and a proactive approach to their online experiences.

C. Addressing online safety and privacy

Ensuring online safety and privacy for youth is paramount in the digital age. Implementing comprehensive strategies that prioritize protection, education, and empowerment is essential. Here are key initiatives to address online safety and privacy concerns among youth:

1. Cybersecurity Education:

- *Integrate:* Incorporate cybersecurity education into school curricula.

- *Empower:* Equip youth with the knowledge to recognize online threats, use secure passwords, and understand the importance of software updates.

2. Privacy Settings Guidance:

- *Educate:* Provide guidance on adjusting privacy settings across various platforms.

- *Empower:* Enable youth to control their online visibility and manage who has access to their personal information.

3. Digital Footprint Awareness:

- *Teach:* Educate youth about the concept of a digital footprint.

- *Highlight:* Emphasize the long-term consequences of online activities and the impact on personal and professional life.

4. Responsible Social Media Use:

- *Educate:* Offer workshops on responsible social media use.

- *Discuss:* Engage in conversations about the potential risks and benefits of sharing personal information online.

5. Reporting Mechanisms Training:

- *Train:* Educate youth on how to use reporting mechanisms on social media platforms.

- *Empower:* Enable them to take proactive steps in reporting inappropriate content, cyberbullying, or suspicious activities.

6. Personal Information Sensitivity:

- *Stress:* Emphasize the sensitivity of personal information.

• *Limit:* Encourage youth to limit the sharing of private details, such as addresses and phone numbers, to trusted individuals.

7. Online Stranger Awareness:

• *Discuss:* Engage in open discussions about online stranger danger.

• *Role-Play:* Conduct role-playing exercises to help youth recognize and respond to potentially unsafe online interactions.

8. Two-Factor Authentication Adoption:

• *Advocate:* Advocate for the use of two-factor authentication.

• *Implement:* Encourage youth to implement this additional layer of security to protect their online accounts.

9. Anti-Phishing Measures:

• *Educate:* Teach youth about phishing and online scams.

• *Verify:* Instill the habit of verifying the authenticity of emails and messages, especially those requesting personal information.

10. Online Gaming Safety:
- *Guidance:* Provide guidelines for safe online gaming.
- *Awareness:* Make youth aware of the potential risks associated with online gaming, including cyberbullying and inappropriate content.
11. Parental Controls Utilization:
- *Inform:* Educate parents on the use of parental control tools.
- *Collaborate:* Encourage collaboration between parents and educators to ensure consistent monitoring of online activities.
12. Digital Privacy Laws Advocacy:
- *Advocate:* Advocate for digital privacy laws that protect youth.
- *Raise Awareness:* Promote awareness of existing regulations and ensure their implementation to safeguard online privacy.
13. Establishing Trust-Based Communication:

- *Foster:* Cultivate trust-based communication between parents, educators, and youth.

- *Encourage:* Encourage youth to openly discuss online safety concerns without fear of judgment.

14. Regular Check-Ins on Online Activities:

- *Implement:* Conduct regular check-ins on youth's online activities.

- *Stay Informed:* Stay informed about the platforms they use, their online friends, and any concerns they may have.

By implementing these comprehensive strategies, educators, parents, and communities can collaboratively create a secure digital environment for youth, empowering them to navigate online spaces with awareness, responsibility, and confidence.

D. Encouraging positive online behaviors

Encouraging positive online behaviors among youth involves fostering a culture of respect, empathy, and responsible digital citizenship. By instilling values that prioritize kindness, collaboration, and constructive engagement, we can shape a positive and inclusive digital environment. Here are key strategies to promote positive online behaviors:

1. Digital Citizenship Education:

- *Integrate:* Incorporate digital citizenship education into school curricula.

- *Empower:* Equip youth with the knowledge and skills to engage responsibly, ethically, and positively online.

2. Online Respect Campaigns:

- *Launch:* Initiate campaigns promoting online respect.

- *Showcase:* Showcase positive examples of individuals using social media to uplift others, fostering a culture of mutual respect.

3. Peer Recognition Programs:

• *Establish:* Create programs recognizing positive online behaviors.

• *Highlight:* Showcase and celebrate individuals who contribute positively to digital communities, encouraging others to follow suit.

4. Empathy-Building Activities:

• *Incorporate:* Integrate empathy-building activities into educational settings.

• *Discuss:* Facilitate discussions on understanding diverse perspectives and the impact of words and actions online.

5. Positive Content Creation Challenges:

• *Initiate:* Launch challenges encouraging positive content creation.

• *Inspire:* Motivate youth to use their creativity to spread messages of kindness, inclusivity, and inspiration.

6. Online Mentorship Programs:

• *Establish:* Create mentorship programs pairing experienced individuals with youth.

• *Guide:* Provide guidance on responsible and positive online behaviors, fostering a sense of community and support.

7. Digital Kindness Pledges:

• *Encourage:* Encourage youth to take digital kindness pledges.

• *Promote:* Share these pledges within schools and communities to create a collective commitment to positive online interactions.

8. Constructive Feedback Culture:

• *Teach:* Educate youth on providing constructive feedback online.

• *Model:* Demonstrate and encourage a culture where feedback is delivered respectfully and helps individuals grow.

9. Positive Role Model Platforms:

• *Showcase:* Highlight platforms that showcase positive online role models.

• *Inspire:* Showcase individuals using social media for uplifting causes, encouraging youth to follow in their footsteps.

10. Online Collaboration Initiatives:
- *Facilitate:* Create opportunities for collaborative online initiatives.
- *Teamwork:* Encourage youth to work together on projects that promote positive values, fostering a sense of teamwork and shared goals.
11. Digital Acts of Kindness Challenges:
- *Launch:* Initiate challenges focused on digital acts of kindness.
- *Celebrate:* Celebrate and amplify instances where youth make positive contributions online, reinforcing a culture of kindness.
12. Positive Content Recognition Awards:
- *Establish:* Create awards recognizing positive online content.
- *Motivate:* Motivate youth to produce content that contributes to a positive and uplifting online environment.
13. Interactive Digital Storytelling:
- *Promote:* Encourage digital storytelling that promotes positive themes.
- *Engagement:* Foster engagement through narratives that inspire, educate, and foster empathy.
14. Celebrating Digital Milestones:
- *Acknowledge:* Acknowledge and celebrate positive digital milestones.
- *Positive Reinforcement:* Provide positive reinforcement for responsible and constructive online behaviors, encouraging continued positive engagement.

By incorporating these strategies into educational programs, community initiatives, and family discussions, we can collectively shape a digital culture that emphasizes positivity, empathy, and responsible citizenship, empowering youth to contribute meaningfully to the online world.

VIII. Future Trends and Challenges

As social media continues to evolve, shaping and being shaped by the youth culture, several trends and challenges emerge on the horizon. Anticipating and addressing these developments is crucial for fostering a positive and responsible digital landscape for the younger generation. Here are the future trends and challenges for social media and youth:

1. *Rise of New Platforms and Features:*

- Trend: The emergence of new social media platforms and innovative features.

- Challenge: Navigating the influx of platforms, each with its unique dynamics, privacy settings, and potential risks.

2. *Augmented Reality (AR) and Virtual Reality (VR) Integration:*

- Trend: Increasing integration of AR and VR technologies into social media experiences.

- Challenge: Balancing the immersive potential of these technologies with concerns about privacy, online safety, and potential addiction.

3. *Influence of Artificial Intelligence (AI) on Content Curation:*

- Trend: AI-driven content curation shaping personalized user experiences.

- Challenge: Mitigating the risk of algorithmic bias, filter bubbles, and the potential impact on youth perspectives and opinions.

4. *Digital Well-Being Initiatives:*

• Trend: Growing emphasis on digital well-being initiatives.

• Challenge: Ensuring the effectiveness of well-being features and interventions, and addressing the mental health impact of social media on youth.

5. *Ephemeral Content and Disappearing Messages:*

• Trend: The continued popularity of ephemeral content and disappearing messages.

• Challenge: Navigating the implications for digital permanence, online reputation management, and responsible sharing.

6. *Increased Integration of E-Commerce:*

• Trend: Social media platforms evolving into e-commerce hubs.

• Challenge: Balancing the opportunities for entrepreneurship with concerns about privacy, security, and the potential for online scams.

7. *Global Collaboration for Online Safety:*

• Trend: Increased global collaboration and efforts to enhance online safety.

• Challenge: Addressing the diverse cultural and legal considerations surrounding online safety, and fostering international cooperation.

8. *Youth Activism and Social Change:*

• Trend: Continued use of social media as a platform for youth activism.

• Challenge: Managing the potential for misinformation, online polarization, and the ethical considerations of activism in the digital age.

9. *Digital Literacy as a Core Skill:*

• Trend: Digital literacy becoming a foundational skill in education.

• Challenge: Ensuring comprehensive and up-to-date digital literacy education that equips youth to navigate evolving online landscapes.

10. *Privacy Concerns and Regulation:*

- Trend: Increasing focus on privacy concerns and potential regulatory measures.

- Challenge: Striking a balance between user privacy, platform innovation, and the need for regulatory frameworks that protect youth.

11. *Gamification of Social Media:*

- Trend: Growing integration of gamification elements to enhance user engagement.

- Challenge: Addressing the potential for addictive behaviors, cyberbullying within gaming communities, and the impact on mental health.

12. *Integration of Mental Health Support Services:*

- Trend: Greater integration of mental health support services within social media platforms.

- Challenge: Ensuring the accessibility, effectiveness, and ethical considerations of mental health resources embedded in digital spaces.

13. *The Role of Augmented Reality in Education:*

- Trend: Increased use of augmented reality for educational purposes.

- Challenge: Navigating the balance between educational innovation and potential distractions, and addressing issues related to accessibility and inclusivity.

14. *Emergence of New Online Subcultures:*

- Trend: The continual emergence of new online subcultures and communities.

- Challenge: Understanding and addressing the dynamics, influences, and potential risks associated with rapidly evolving digital subcultures.

As social media and youth culture continue to intertwine, stakeholders must remain vigilant, adaptable, and collaborative to proactively address emerging trends and challenges. This ongoing effort ensures that the digital landscape remains a positive and empowering space for the youth of tomorrow.

A. Emerging social media platforms and technologies

The social media landscape is continually evolving, introducing new platforms and technologies that redefine the way youth engage, communicate, and express themselves online. These emerging trends play a pivotal role in shaping the digital experiences of the younger generation. Here are some noteworthy emerging social media platforms and technologies:

1. *TikTok:*

- Overview: TikTok has surged in popularity, especially among younger users, with its short-form videos set to music. It emphasizes creativity, trends, and user-generated content.

- Impact: Redefining content creation by encouraging spontaneous and authentic expression, fostering a new wave of influencers and trends.

2. *Clubhouse:*

- Overview: Clubhouse is an audio-based social networking app where users can join virtual rooms for live, unscripted conversations on various topics.

- Impact: Transforming the way users engage in real-time discussions, providing a platform for authentic conversations and knowledge sharing.

3. *Snapchat Spotlight:*

• Overview: Snapchat Spotlight is a feature within the Snapchat app that showcases user-generated short-form videos. It is designed to highlight engaging content and reward creators.

• Impact: Competing with other short-form video platforms, Spotlight incentivizes content creation and introduces a new way for users to discover and engage with content.

4. *NFTs in Social Media:*

• Overview: Non-fungible tokens (NFTs) are unique digital assets verified through blockchain technology. Social media platforms are exploring ways to integrate NFTs, allowing users to own and trade digital content.

• Impact: Redefining ownership of digital creations, enabling creators to monetize their work and fostering a new era of digital collectibles.

5. *Instagram Reels:*

• Overview: Instagram Reels is a short-form video feature within the Instagram platform, similar to TikTok. It allows users to create and discover entertaining content.

• Impact: Instagram's response to the popularity of TikTok, providing users with another avenue for creative expression and fostering a competitive landscape for short-form videos.

6. *Virtual Influencers:*

• Overview: Virtual influencers are computer-generated characters with their own personalities and social media presence. Brands are increasingly using them for marketing and collaborations.

• Impact: Blurring the lines between real and virtual, virtual influencers raise questions about authenticity, representation, and the future of influencer marketing.

7. *Spatial Audio and 3D Audio:*

• Overview: Social media platforms are exploring spatial audio and 3D audio features to enhance the immersive experience of virtual conversations and content.

• Impact: Transforming the auditory experience in virtual spaces, providing users with a more immersive and engaging way to connect.

8. *Meta-verse Platforms:*

• Overview: Metaverse platforms are immersive virtual spaces where users can interact with a digital environment. Social media is exploring metaverse concepts for enhanced user experiences.

• Impact: Redefining the concept of online social interaction, offering users virtual spaces to connect, socialize, and engage in diverse activities beyond traditional platforms.

9. *AI-Generated Content:*

• Overview: Artificial intelligence (AI) is increasingly being used to generate content, including images, videos, and written text. Social media platforms are exploring AI-generated features.

• Impact: Raising questions about the authenticity of content, potential misuse, and the role of AI in shaping digital creativity.

10. *Livestream Shopping:*
- Overview: Livestream shopping integrates e-commerce with live video streaming, allowing users to shop in real-time while interacting with hosts and other viewers.

- Impact: Revolutionizing the online shopping experience, creating a dynamic and interactive platform for brands and influencers to showcase products.

11. *Blockchain-Based Social Networks:*

- Overview: Blockchain technology is being explored for creating decentralized social networks. These platforms aim to give users more control over their data and privacy.

- Impact: Addressing concerns about data ownership and privacy, blockchain-based social networks aim to provide users with greater transparency and control.

As these emerging platforms and technologies gain traction, they significantly influence how youth interact with the digital world. It's crucial for educators, parents, and stakeholders to stay informed and guide youth in navigating these evolving digital landscapes responsibly.

B. Potential societal and cultural shifts

The dynamic nature of social media and its influence on youth culture can lead to significant societal and cultural shifts. As platforms evolve and new trends emerge, they have the potential to reshape the way youth interact, communicate, and perceive the world. Here are some potential societal and cultural shifts influenced by the evolving landscape of social media:

1. *Digital Identity and Self-Expression:*

- Potential Shift: A shift towards more nuanced and diverse forms of digital self-expression.

- Impact: Youth may increasingly embrace multifaceted digital identities, challenging traditional norms and fostering a more inclusive online culture.

2. *Global Connectivity and Awareness:*

- Potential Shift: Increased global awareness and interconnectedness among youth.

- Impact: Social media's role in facilitating cross-cultural interactions may lead to a generation with a broader worldview, enhanced empathy, and a greater sense of global citizenship.

3. *Influence on Cultural Trends:*

- Potential Shift: Acceleration of cultural trends driven by social media.

- Impact: Cultural shifts, including fashion, language, and social norms, may occur more rapidly as trends disseminate quickly through online platforms.

4. *Redefining Social Norms:*

- Potential Shift: Redefinition of social norms influenced by online communities.

- Impact: Social media movements may challenge and reshape traditional norms, fostering a more inclusive and accepting societal landscape.

5. *New Forms of Activism:*

- Potential Shift: Emergence of new forms of online activism and advocacy.

- Impact: Social media platforms may continue to serve as catalysts for social change, providing youth with powerful tools to advocate for causes and mobilize collective action.

6. *Evolving Notions of Privacy:*

- Potential Shift: Evolving attitudes towards privacy in the digital age.

• Impact: Youth may become more conscious of their online presence, leading to a reevaluation of what is considered private and public in the digital sphere.

7. *Impact on Mental Health Conversations:*

• Potential Shift: Increased openness and conversations around mental health.

• Impact: Social media's role in destigmatizing mental health issues may contribute to a more supportive and understanding society, especially among younger generations.

8. *Digital Activism and Civic Engagement:*

• Potential Shift: Growing engagement in digital activism and civic participation.

• Impact: Social media may encourage youth to actively participate in political and social issues, influencing the landscape of civic engagement.

9. *Changes in Communication Styles:*

• Potential Shift: Transformation in communication styles influenced by digital platforms.

• Impact: The use of emojis, memes, and short-form content may reshape how youth communicate, fostering new and creative means of expression.

10. *Virtual Collaboration and Remote Work Norms:*
- Potential Shift: Integration of virtual collaboration norms into professional life.
- Impact: The rise of remote work and online collaboration during youth formative years may influence future professional expectations and norms.
11. *Reimagining Education and Learning:*

- Potential Shift: Transformation of education models influenced by online learning.

- Impact: The digital-native generation may prioritize flexible, technology-driven learning approaches, influencing the future of education.

12. *Shifts in Consumer Behavior:*

- Potential Shift: Changes in consumer behavior driven by social media influencers.

- Impact: Youth may increasingly turn to influencers for product recommendations, challenging traditional advertising models and shaping consumer trends.

13. *Erosion of Traditional Media Gatekeepers:*

- Potential Shift: Diminishing influence of traditional media gatekeepers.

- Impact: Social media's democratization of information may lead to a decline in the influence of traditional media outlets, with youth relying more on peer-driven content.

14. *Cultural Sensitivity and Inclusivity:*

- Potential Shift: Heightened awareness and emphasis on cultural sensitivity.

- Impact: Youth may prioritize inclusivity and cultural awareness, driven by exposure to diverse perspectives on social media.

Navigating these potential shifts requires a proactive approach from educators, parents, and society at large. By understanding and actively engaging with the evolving digital landscape, stakeholders can support youth in developing a positive and responsible relationship with social media, contributing to the formation of a more inclusive and informed generation.

C. Ethical considerations and regulation

As social media continues to shape the experiences of youth, ethical considerations and regulatory frameworks play a crucial role in ensuring a safe and positive digital environment. Addressing these aspects is essential to protect the well-being of young individuals. Here are key considerations related to ethics and regulation in the context of social media and youth:

1. *Privacy Concerns:*

• Ethical Consideration: Respecting the privacy rights of youth and ensuring their personal information is handled responsibly.

• Regulation: Advocating for and enforcing robust data protection laws that safeguard the privacy of young users.

2. *Age Verification and COPPA Compliance:*

• Ethical Consideration: Ensuring platforms comply with age restrictions and provide a safe space for age-appropriate content.

• Regulation: Enforcing laws like the Children's Online Privacy Protection Act (COPPA) to protect children under the age of 13 online.

3. *Algorithmic Transparency:*

• Ethical Consideration: Transparent disclosure of algorithms to prevent unintended consequences, bias, and manipulation.

• Regulation: Advocating for regulations that require social media platforms to be transparent about their algorithms and how they impact user experiences.

4. *Combatting Cyberbullying:*

• Ethical Consideration: Taking proactive measures to prevent and address cyberbullying, fostering a culture of kindness and respect.

• Regulation: Implementing and enforcing anti-cyberbullying laws that hold individuals accountable for harmful online behavior.

5. *Preventing Exploitative Advertising:*

• Ethical Consideration: Protecting youth from manipulative or exploitative advertising practices.

- Regulation: Advocating for regulations that limit targeted advertising to youth and ensure transparency in ad practices.

6. *Mental Health Safeguards:*

- Ethical Consideration: Prioritizing the mental health of youth by minimizing harmful content and fostering positive online experiences.

- Regulation: Implementing guidelines and regulations that promote mental health safeguards on social media platforms.

7. *Addressing Online Predators:*

- Ethical Consideration: Vigilance in protecting youth from online predators and ensuring a secure digital environment.

- Regulation: Enforcing laws that address online grooming, exploitation, and the protection of minors from predatory behavior.

8. *Countering Disinformation:*

- Ethical Consideration: Combating the spread of misinformation to protect the integrity of information consumed by youth.

- Regulation: Developing and enforcing regulations that hold platforms accountable for addressing and curbing the dissemination of false information.

9. *Inclusive Design and Accessibility:*

- Ethical Consideration: Designing platforms that are inclusive and accessible to all, regardless of physical abilities or differences.

- Regulation: Advocating for and enforcing regulations that promote inclusive design practices in social media platforms.

10. *Digital Literacy Education:*

- Ethical Consideration: Empowering youth with digital literacy skills to navigate the online world responsibly.

- Regulation: Advocating for the integration of digital literacy education into formal curricula to equip youth with the tools to critically assess online content.

11. *Combatting Online Hate Speech:*

- Ethical Consideration: Taking a strong stance against online hate speech and fostering an environment of tolerance and understanding.

- Regulation: Enforcing laws that address and penalize hate speech on social media platforms.

12. *User Consent and Control:*

- Ethical Consideration: Ensuring users, including youth, have informed consent and control over their data.

- Regulation: Advocating for regulations that strengthen user control and consent mechanisms, particularly for younger users.

13. *Balancing Freedom of Expression:*

- Ethical Consideration: Balancing the right to freedom of expression with the responsibility to prevent harm and protect vulnerable users.

- Regulation: Developing frameworks that strike a balance between free expression and preventing harm, particularly when it comes to youth safety.

14. *Community Moderation and Accountability:*

- Ethical Consideration: Implementing fair and transparent community moderation practices that hold users accountable for their behavior.

- Regulation: Advocating for regulations that ensure social media platforms have effective moderation processes in place to address harmful content.

Ethical considerations and regulatory measures should work hand-in-hand to create a digital landscape that prioritizes the well-being, safety, and rights of youth. This requires ongoing collaboration between policymakers, tech industry stakeholders, educators, parents, and youth themselves to address emerging challenges and foster a responsible and positive online environment.

Conclusion

The intersection of social media and youth has given rise to a transformative digital era, shaping how young individuals communicate, express themselves, and navigate the complexities of the online world. This book has delved into the multifaceted impact of social media on youth, exploring its positive aspects, challenges, and the ethical considerations and regulations essential for safeguarding their well-being.

In tracing the evolution of social media, we witnessed the emergence of platforms like TikTok, Clubhouse, and Instagram Reels, each influencing how youth create and consume content. These platforms not only reflect the trends of the moment but also hold the potential to redefine cultural norms, communication styles, and global connectivity.

Amid the positive impacts explored, such as enhanced social connectivity, educational opportunities, and avenues for creative expression, we acknowledged the darker side. The negative impacts, including mental health challenges, cyberbullying, privacy concerns, and misinformation, underscore the need for a vigilant approach in guiding youth through the digital landscape.

Ethical considerations and regulatory frameworks emerged as crucial components in ensuring a safe and responsible online environment for youth. Balancing the right to privacy, freedom of expression, and the prevention of harm requires ongoing collaboration between stakeholders. Whether addressing issues of privacy, combatting cyberbullying, or promoting digital literacy, ethical considerations and regulations must evolve alongside the dynamic social media landscape.

As we contemplate the future, the potential societal and cultural shifts fueled by social media paint a picture of a generation deeply interconnected, globally aware, and influential in shaping cultural norms. The responsibility lies with educators, parents, policymakers, and tech industry leaders to guide youth through this transformative journey, fostering a positive, inclusive, and empowering digital space.

In conclusion, the impact of social media on youth is profound, multifaceted, and continually evolving. By embracing the positive aspects, proactively addressing challenges, and upholding ethical standards, we can collectively shape a digital landscape that empowers youth to thrive in the ever-changing social media landscape while preserving their well-being and fostering a future where technology enhances rather than hinders their growth.

A. Summarizing the impact of social media on youth

The impact of social media on youth is a nuanced interplay of positive and negative influences that shape their experiences in the digital age. Here's a concise summary:

Positive Impacts:

Social Connectivity and Peer Relationships:

• Social media enhances youth connectivity, fostering peer relationships and a sense of belonging.

Information and Educational Opportunities:

• It provides a vast platform for information, educational resources, and learning opportunities.

Creative Expression and Self-Identity:

• Social media empowers youth to express creativity, explore their identities, and share diverse perspectives.

Entrepreneurship and Online Opportunities:

• It opens avenues for entrepreneurship, allowing youth to showcase talents and explore online opportunities.

Activism and Social Change:

- Social media serves as a catalyst for youth activism, enabling them to engage in social issues and drive change.

Negative Impacts:

Mental Health Challenges and Social Comparison:

- Youth face mental health challenges, including social comparison and unrealistic standards perpetuated on social media.

Cyberbullying and Online Harassment:

- Online platforms expose youth to cyberbullying and harassment, impacting their well-being.

Privacy Concerns and Digital Footprints:

- There are concerns about privacy and the lasting impact of digital footprints on youth.

Addiction and Screen Time:

- Excessive screen time and social media use contribute to addiction and potential negative effects on mental health.

Fake News and Misinformation:

- Youth are susceptible to misinformation, influencing their perspectives and contributing to the spread of fake news.

In navigating this complex landscape, addressing ethical considerations, implementing regulations, and promoting digital literacy are essential for fostering a positive and responsible online environment for youth. The evolving trends, potential cultural shifts, and ongoing challenges underscore the importance of proactive engagement from educators, parents, policymakers, and technology platforms in guiding youth through the dynamic world of social media.

B. The importance of responsible use and education

Responsible use of social media and comprehensive education on digital literacy are paramount in safeguarding the well-being of youth in the digital age. Here's a summary of their significance:

Mitigating Negative Impacts:

● Responsible use helps mitigate the negative impacts of social media, such as cyberbullying, mental health challenges, and exposure to harmful content.

Empowering Critical Thinking:

● Digital literacy education empowers youth with critical thinking skills, enabling them to discern credible information from misinformation.

Privacy Protection:

● Responsible use involves safeguarding personal information, and digital literacy education helps youth understand the importance of privacy settings and the potential consequences of oversharing.

Preventing Online Harassment:

● Education on responsible digital citizenship contributes to a culture of respect, preventing online harassment and promoting positive online interactions.

Promoting Ethical Behavior:

• Digital literacy education instills ethical behavior online, fostering a sense of responsibility and accountability for one's actions in the digital space.

Navigating Social Pressures:

• Responsible use encourages youth to manage social pressures, promoting self-esteem and a healthy digital self-identity.

Balancing Screen Time:

• Education on responsible screen time management helps youth strike a balance between online and offline activities, mitigating the risks of addiction and excessive exposure.

Addressing Online Safety:

• Responsible use involves being vigilant about online safety, and digital literacy education equips youth with the knowledge to identify and report potential risks.

Enhancing Career Opportunities:

• Digital literacy skills contribute to enhanced career opportunities, as youth learn to leverage social media for professional networking and personal branding.

Creating Informed Citizens:

• Digital literacy education creates informed citizens who can critically engage with information, contribute positively to online discussions, and participate in democratic processes.

Preventing Exploitation:

• Responsible use helps prevent the exploitation of youth through online platforms, and digital literacy education raises awareness about potential risks and scams.

Building Resilience:

• Education on responsible digital behavior builds resilience in youth, helping them navigate challenges, setbacks, and online pressures.

Encouraging Positive Online Behaviors:

• Responsible use fosters positive online behaviors, while digital literacy education encourages youth to contribute to a culture of kindness, respect, and inclusivity.

In essence, the combination of responsible social media use and robust digital literacy education forms a powerful strategy to empower youth in the digital landscape. By instilling ethical principles, critical thinking skills, and a sense of responsibility, we can guide the younger generation towards a positive and informed engagement with social media, ensuring that it becomes a tool for growth, connection, and learning rather than a source of harm and misinformation.

C. Looking ahead to the future of social media and its role in youth culture

As we peer into the future, the trajectory of social media holds immense potential to shape the youth culture in profound ways. Here's a glimpse of what lies ahead:

Innovative Platforms and Features:

• *Anticipated Trend:* Continued emergence of innovative platforms and features.

- *Impact:* New platforms and features will redefine how youth engage with content, fostering creativity and driving cultural trends.

Augmented Reality (AR) and Virtual Reality (VR):

- *Anticipated Trend:* Increased integration of AR and VR technologies.

- *Impact:* AR and VR will transform social media experiences, offering immersive interactions and innovative forms of content creation.

Artificial Intelligence (AI) in Content Curation:

- *Anticipated Trend:* Growing influence of AI-driven content curation.

- *Impact:* Personalized user experiences will evolve, raising questions about algorithmic transparency, bias, and the ethical use of AI.

Digital Well-Being Initiatives:

- *Anticipated Trend:* Heightened focus on digital well-being initiatives.

- *Impact:* Platforms will increasingly prioritize features that promote mental health, balance, and positive user experiences.

Ephemeral Content and Disappearing Messages:

- *Anticipated Trend:* Continued popularity of ephemeral content.

- *Impact:* Youth culture will adapt to a dynamic content-sharing landscape, where impermanence shapes online communication.

Increased Integration of E-Commerce:

- *Anticipated Trend:* Social media evolving into prominent e-commerce hubs.

- *Impact:* Youth will explore entrepreneurial opportunities and redefine how they engage with brands and online shopping.

Global Collaboration for Online Safety:

- *Anticipated Trend:* Growing global collaboration for enhanced online safety.

- *Impact:* Cross-border initiatives will shape international standards, fostering safer digital spaces for youth.

Youth Activism and Social Change:

- *Anticipated Trend:* Continued use of social media for youth activism.

- *Impact:* Social media will remain a powerful tool for mobilizing young activists and driving positive social change.

Digital Literacy as a Core Skill:

- *Anticipated Trend:* Digital literacy becoming a foundational skill.

- *Impact:* Education systems will adapt to equip youth with the critical skills needed to navigate an ever-evolving digital landscape.

Privacy Concerns and Regulation:

- *Anticipated Trend:* Increasing scrutiny on privacy concerns and potential regulation.

- *Impact:* Stricter regulations may reshape how platforms handle user data, impacting youth's online privacy and security.

Gamification of Social Media:

• *Anticipated Trend:* Growing integration of gamification elements.

• *Impact:* Platforms will leverage gamification to enhance engagement, posing challenges related to addiction and responsible use.

Integration of Mental Health Support Services:

• *Anticipated Trend:* Greater integration of mental health support services.

• *Impact:* Social media platforms will actively contribute to mental health awareness and provide accessible resources for users.

The Role of Augmented Reality in Education:

• *Anticipated Trend:* Increased use of augmented reality for educational purposes.

• *Impact:* Learning experiences will become more interactive and immersive, revolutionizing education for the digital-native generation.

Emergence of New Online Subcultures:

• *Anticipated Trend:* Continuous emergence of new online subcultures.

• *Impact:* Youth will define and participate in evolving digital subcultures, shaping the online narrative in unique and diverse ways.

In navigating this future, it is imperative for stakeholders to remain adaptive, proactive, and collaborative. By understanding and embracing the evolving trends, educators, parents, policymakers, and technology leaders can guide youth towards responsible, positive, and empowering engagements with the ever-evolving landscape of social media.

Also by Swatantra Bahadur

Breaking Barriers: LGBTQ Rights and Social Justice
Blossom with confidence
"Depression: A Roller Coaster Ride"
Finding Your Voice
Rahul Gandhi: The Untold Story
100 Aspects on Nature
Love By An Introvert
Man Of Golden India "Narendra Modi"
India " Unity lies in Diversity"
Indian's Heritage of Kashi "Varanasi"
"The Power of Voice: Lawyer in a Black Coat"
Shri Ram Janmabhumi "Ayodhya"
Social Media and Youth: Navigating the Digital Landscape

About the Author

Instagram Id - swatantrabahadur15